Communications in Computer and Information Science 1200

Commenced Publication in 2007
Founding and Former Series Editors:
Simone Diniz Junqueira Barbosa, Phoebe Chen, Alfredo Cuzzocrea,
Xiaoyong Du, Orhun Kara, Ting Liu, Krishna M. Sivalingam,
Dominik Ślęzak, Takashi Washio, Xiaokang Yang, and Junsong Yuan

Editorial Board Members

More information about this series at http://www.springer.com/series/7899

Franco Cicirelli · Antonio Guerrieri ·
Clara Pizzuti · Annalisa Socievole ·
Giandomenico Spezzano ·
Andrea Vinci (Eds.)

Artificial Life and Evolutionary Computation

14th Italian Workshop, WIVACE 2019
Rende, Italy, September 18–20, 2019
Revised Selected Papers

Springer

Editors
Franco Cicirelli (iD)
ICAR-CNR
Rende, Italy

Antonio Guerrieri (iD)
ICAR-CNR
Rende, Italy

Clara Pizzuti (iD)
ICAR-CNR
Rende, Italy

Annalisa Socievole (iD)
ICAR-CNR
Rende, Italy

Giandomenico Spezzano (iD)
ICAR-CNR
Rende, Italy

Andrea Vinci (iD)
ICAR-CNR
Rende, Italy

ISSN 1865-0929   ISSN 1865-0937 (electronic)
Communications in Computer and Information Science
ISBN 978-3-030-45015-1   ISBN 978-3-030-45016-8 (eBook)
https://doi.org/10.1007/978-3-030-45016-8

This Springer imprint is published by the registered company Springer Nature Switzerland AG
The registered company address is: Gewerbestrasse 11, 6330 Cham, Switzerland

Preface

This volume of *Communication in Computer and Information Science* contains the proceedings of WIVACE 2019, the XIV Workshop on Artificial Life and Evolutionary Computation. The event was successfully held on the Campus of the University of Calabria, Italy, during September 18–20, 2019. WIVACE aims to bring together researchers working in the field of artificial life and evolutionary computation to present and share their research in a multidisciplinary context. This year a special session focusing on cognitive systems and their applications was organized. The workshop provided a platform for academic and researchers to present papers addressing fundamental and practical issues in cognitive computing. Nowadays, we are crossing a new frontier in the evolution of computing and entering the era of cognitive systems. Whereas in the programmable era computers essentially processed a series of "if-then-else" formulas, cognitive systems learn, adapt, and ultimately hypothesize and suggest answers. Cognitive systems promise to tackle complexity and assist people and organizations in making decisions in different application fields such as healthcare, energy, retail, smart machines, smart buildings and cities, environmental modeling, education, banking, food science and agriculture, and climate change.

WIVACE 2019 received 31 submissions in total, out of which 16 were selected for publication in an extended version in this proceedings volume, after a single-blind review round performed by at least three Program Committee members, with an acceptance rate of 54% of the original submissions. Submissions and participants of WIVACE 2019 came from nine different countries, making WIVACE an increasingly international event, despite its origins as an Italian workshop. Following this ever-increasing international spirit, the future WIVACE edition will be held in Switzerland.

Many people contributed to this successful edition. The editors wish to express their sincere gratitude to all who supported this venture. In particular, we wish to thank all the authors who contributed to this volume. We would also like to thank reviewers who, as members of the Program Committee, not only assessed papers but also acted as session chairmen during the workshop.

We want to give special thanks to the invited speakers of both WIVACE 2019 and the special session on cognitive computing which, during the workshop, gave three very interesting and inspiring talks: Prof. Heiko Hamann from the University of Lübeck, Institute of Computer Engineering, Germany; Dr. Yulia Sandamirskaya from the Institute of Neuroinformatics, Neuroscience Center, Zurich University and ETH Zurich, Switzerland; and Prof. Giulio Sandini from Central Research Labs Genova - Italian Institute of Technology and University of Genoa, Italy.

Our gratitude also goes to SenSysCal, a spin-off of the University of Calabria, for giving its support to the workshop administration.

Finally, we acknowledge the helpful advice of the staff at Springer, who provided professional support through all the phases that led to this volume.

September 2019

Franco Cicirelli
Antonio Guerrieri
Clara Pizzuti
Annalisa Socievole
Giandomenico Spezzano
Andrea Vinci

Organization

General Chairs

Clara Pizzuti	ICAR-CNR, Italy
Giandomenico Spezzano	ICAR-CNR, Italy

Local Chairs

Franco Cicirelli	ICAR-CNR, Italy
Antonio Guerrieri	ICAR-CNR, Italy
Annalisa Socievole	ICAR-CNR, Italy
Andrea Vinci	ICAR-CNR, Italy

Special Session Chairs

Giandomenico Spezzano	ICAR-CNR, Italy
Edoardo Serra	Boise State University, USA

Program Committee

Michele Amoretti	University of Parma, Italy
Marco Baioletti	University of Perugia, Italy
Vito Antonio Bevilacqua	Politecnico di Bari, Italy
Leonardo Bich	Universidad del Pais Vasco (UPV/EHU), Spain
Eleonora Bilotta	University of Calabria, Italy
Leonardo Bocchi	University of Florence, Italy
Michele Braccini	University of Bologna, Italy
Marcello Antonio Budroni	University of Sassari, Italy
Stefano Cagnoni	University of Parma, Italy
Angelo Cangelosi	University of Plymouth, UK
Giulio Caravagna	University of Milan-Bicocca, Italy
Timoteo Carletti	University of Namur, Belgium
Antonio Chella	University of Palermo, Italy
Antonio Della Cioppa	University of Salerno, Italy
Maria Pia Fanti	Politecnico di Bari, Italy
Francesco Fontanella	University of Cassino e del Lazio Meridionale, Italy
Salvatore Gaglio	University of Palermo, Italy
Luigi Gallo	ICAR-CNR, Italy
Mario Giacobini	University of Turin, Italy
Alex Graudenzi	University of Milan-Bicocca, Italy
Gianluigi Greco	University of Calabria, Italy
Giovanni Iacca	University of Trento, Italy

Contents

Towards an Assistive Social Robot Interacting with Human Patient to Establish a Mutual Affective Support

Ignazio Infantino(✉) and Alberto Machí

Istituto di Calcolo e Reti ad Alte Prestazioni, Consiglio Nazionale delle Ricerche,
ICAR-CNR, via Ugo La Malfa 153, 90146 Palermo, Italy
{ignazio.infantino,alberto.machi}@cnr.it
http://www.icar.cnr.it

Abstract. The paper describes an architecture for an assistive robot acting in a domestic environment aiming to establish a robust affective and emotional relationship with the patient during rehabilitation at home. The robot has the aim to support the patient in the therapy, to monitor the patient health state, to give affective support increasing the motivation of the human in a period of two or three weeks. The affective based relationship will arise from an interaction based on natural language verbal interaction, on the acquisition of data and vital parameters by environmental and wearable sensors, on a robust human perception using the perceptive robot capabilities. An important issue is that robot activity should be understandable by the human: the proposed architecture enables the robot to express its (emotional) state, its planes, and its interpretation of perceptual data. The implicit goal is to obtain a human emotional involvement that causes the patient to "take care of" its artificial assistant, trying to satisfy the robot's expectation and motivation. The paper describes the proposed layered architecture (including modules components responsible for events and contexts detection, planning, complex verbal interaction, and artificial motivation, use cases, design patterns and control policies), discusses modeling use cases, and reports preliminary experimentation performed by simulation.

Keywords: Social robots · Human-robot interaction · Assistive robotics

1 Introduction

The proposed research deal with two inter-related aspects of future social robots: the real affective support of the human companion over long periods (years or decades); the robot-ethic substrate that regulates the relationship between human and machine according to a set of given social laws but taking into account also human behavior and character. The robot will be a motivator, an entertainer, an advisor, and in one word, a friend. The research will involve: to

© Springer Nature Switzerland AG 2020
F. Cicirelli et al. (Eds.): WIVACE 2019, CCIS 1200, pp. 1–6, 2020.
https://doi.org/10.1007/978-3-030-45016-8_1

design of new improved sensors (in the environment, wearable, or on the robot) to catch human mood signals; not only facial expressions and postures but also oriented to recognize relevant activity patterns; to process multiple sensor data to create a knowledge base storing human social behavior (habits, mood tendencies, character, preferences); the human-robot relationship will evolve according to not only to assure human satisfaction but also to guarantee a sufficient robot motivation to continue its service; in this sense, human will take care of its artificial companion understanding its expectations and desiderata; improved verbal interaction skills that depend on emotions, mood, experience, motivation (both of the robot and the human). In the past decade, the research in medical and health-care robotics has produced various systems working in real scenarios. Among the proposals of a robot control architecture, in [3] the proposed solution includes sustaining user's motivation and engagement for multiple therapeutic scenarios. Its relevant key specifications are: human-robot personality adaptation, users profile and affect influencing behavior generation and realization, platform-independent behavior, and supervised-autonomy. In [7] in addition to providing mechanical/physical assistance in rehabilitation, robots can also provide personalized monitoring, motivation, and coaching. Event-driven architectures and active database systems need the definition of a structure of active rules and their interaction with the planner. For instance, [6] proposes the ECA (Event condition action) as a short cut to represent such a structure. Regarding the monitoring of the emotional state of the patient, in [9] physiological signals such as heart rate and galvanic skin response allow the system an accurate patient emotional state classification in interaction with nursing robots during medical service algorithms based on Wavelet Analysis and statistics-based feature selection process the signals. Another important domain to consider is Ambient-Assisted Living, given that the developed tools aim to establish a strong social interaction with humans [8] based on ambient intelligence paradigm. A well-known example is GiraffPlus [4] that combines social interaction and long term monitoring. This system has an easy communication tool, the possibility to have meaningful and personalized information about what has happened in the home, the collection and tracking of physiological parameters, and the raising of alarms and warnings in case of need. Another exciting robot companion is described in [5], where everyday conditions in private apartments are the constraints taken into account for domestic health assistance. Innovative human-machine interactions will be characterized by pervasive, unobtrusive, and anticipatory communications, as discussed in the Survey on Ambient Intelligence in Healthcare [1]. Finally, architectures could be consistent with the concepts of Web-of-Things (WoT) by comprising the physical sensors layer linked to a health Web-portal layer via a given network infrastructure. For instance, see the e-Health platform for integrated diabetes care management presented in [2]. Moreover, healthcare cyber-physical systems could be supported by cloud and big data services [10].

2 The Proposed Architecture

The paper presents the current status of a research aimed to define design patterns and policies for a robot management architecture allowing the robot to express its (emotional) state, its planes, and its interpretation of perceptual data. The implicit goal is to obtain a human emotional involvement that causes the patient to "take care of" its artificial assistant, trying to satisfy the robot's expectation and motivation. The paper describes the proposed architecture (including modules responsible for events and contexts detection, planning, and artificial motivation), discusses a real use case, and reports preliminary experimentation. The architecture consists of a set of controller processes and generic support services for complex dialogue sessions considering the semantic context and behavioral patterns. To express behavior in the dialogue, the robotic assistant could exhibit a sort of self-awareness, to assure the explanation of motivations, and to show affective attitudes. Referring to the abstract OSI (Open Standard Interconnection) programming model, such architecture defines archetypes of controller processes, cooperation protocols, OSI 5 management policies for the application, responsible for maintaining semantic and operational consistency at the exchange session level of robot-patient messages. For the management of the communication infrastructure based on messages/events, we assume the use of the frameworks and services provided by WebSockets (see www.websocket.org) and the OSI level 4 of ROS (Robot Operating System, see wiki.ros.org). In the considered scenario, the assistant robot has the primary function of entertainer and motivator the patient to keep his behavior punctually compliant with the prescriptions of the therapeutic protocol, while a network of wearable sensors assures the monitoring of physiological parameters. The robot therefore acts as a personal assistant and home automation for most of the time. The function of monitoring the state of health is outside of its features and level of consciousness, except through the execution of health tasks expressly provided for in the treatment plan for alert activation. Based on a dynamically modifiable plan (OSI level 7) updated by an external intelligent component, the robot activates knowledge and health functions provided by the therapeutic protocol. Tasks such as notifications, reminders, interviews, the log of verbal expressions and mood, could be executed both according to the timing provided by therapeutic planning, and/or following an oral request for assistance of the patient. In the dialogue, the assistant expresses self-awareness of his internal and environmental operational state and a variable emotional attitude towards the interlocutor, defined by an external ethical control component. In the scenario described above, there are several cooperating actors on whom the intelligence of the system is distributed: Patient, Verbal Robotic Assistant, Interaction Manager, Situation Monitor (environmental, semantic, operational, emotional), Biometric Sensor Manager, Therapy Controller. The patient interacts with the robot requesting information or activities and responding to interviews, requests, and status notifications. The dialogue assistant, synchronized by the interaction manager and activated by the therapy controller or by the patient, carries out verbal interaction sessions (also with gestures) modulated according to behavioral patterns (attitudes) chosen by the planner in function of therapeutic compliance status indicators maintained by the Therapy Controller and the patient's mood evaluated by the Monitor. The Interaction Controller manages the preconditions for interaction (engagement) and its physicality (sampling of input audio, TTS, STT, synchronization with motor functions) and non-verbal expression of attitudes and feelings (gesture). Figure 1 shows the relevant software components of the architecture. In particular, at OSI level 4, it includes the bridge between ROS nodes and the event-based framework by WebSocket. The dialogue manager that uses two Watson (see www.ibm.com/watson/developer/) cognitive assistant services (one generic, and the other specialized on therapy protocol), a gesture database (e.g., by MongoDB), the state monitor, and the activity manager (scheduler and dispatcher) are at OSI level 5. The Interactor is at OSI level 6, and the Health Manager and Planner are at OSI level 7.

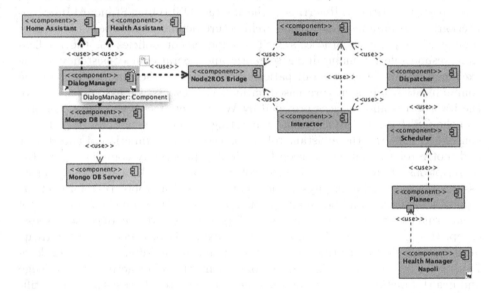

Fig. 1. The proposed software architecture for the assistive social robot combining cognitive capabilities of Watson based verbal interaction services and ROS nodes. ROS modules manage the robot actions by an event-based planning.

3   Scenario

The proposed architecture will be deployed on a humanoid robot for monitoring postoperative home cardiac therapy. The home automation assistant functions considered are inspired by [11], and the services of health assistant are inspired by [12]. The robot functionalities concern the entertainment, the research, and supply of information, the activation of environmental controls and telematics, the monitoring of biomedical parameters, the request to answer a questionnaire, the assistance in the execution of activities based on a daily agenda and a therapy plan. The interaction takes place both through explicit exchange sessions (during a human-robot engagement) by sentences in natural language (Italian) and through the exchange with and between control processes of management messages. During the dialogue, the patient and the assistant negotiate the control of the session. An entertainment session can be interrupted for the execution of a non-deferrable health task. A deferrable system notification can be postponed to the end of a dialogue session. Verbal interactions aims to check therapy adherence and to monitor mood of the patient. By processing the human answers, the robot define and modulate its behavior as interactive assistant, adapting both to the user expectation and the health operators desiderata.

Acknowledgments. The work has been supported by the Italian project "AMICO - Medical Assistence In COntextual awareness" , ARS 0100900 - CUP B46G18000390005, Decreto GU n.267/16-11-2018, PON R&I 2014–2020.

References

1. Acampora, G., Cook, D.J., Rashidi, P., Vasilakos, A.V.: A survey on ambient intelligence in healthcare. Proc. IEEE **101**(12), 2470–2494 (2013). https://doi.org/10.1109/JPROC.2013.2262913
2. Al-Taee, M.A., Sungoor, A.H., Abood, S.N., Philip, N.Y.: Web-of-things inspired e-health platform for integrated diabetes care management. In: 2013 IEEE Jordan Conference on Applied Electrical Engineering and Computing Technologies (AEECT), pp. 1–6 (2013). https://doi.org/10.1109/AEECT.2013.6716427
3. Cao, H.L., Esteban, P.G., Beir, A.D., Simut, R., de Perre, G.V., Vanderborght, B.: A platform-independent robot control architecture for multiple therapeutic scenarios. CoRR abs/1607.04971 (2016)
4. Coradeschi, S., et al.: Giraffplus: Combining social interaction and long term monitoring for promoting independent living. In: 2013 6th International Conference on Human System Interactions (HSI), pp. 578–585 (2013). https://doi.org/10.1109/HSI.2013.6577883
5. Gross, H., et al.: Robot companion for domestic health assistance: Implementation, test and case study under everyday conditions in private apartments. In: 2015 IEEE/RSJ International Conference on Intelligent Robots and Systems (IROS), pp. 5992–5999 (2015)
6. Hossain, M.A., Ahmed, D.T.: Virtual caregiver: An ambient-aware elderly monitoring system. IEEE Tran. Inf. Technol. Biomed. **16**(6), 1024–1031 (2012). https://doi.org/10.1109/TITB.2012.2203313
7. Okamura, A.M., Mataric, M.J., Christensen, H.I.: Medical and health-care robotics. IEEE Robot. Autom. Mag. **17**(3), 26–37 (2010). https://doi.org/10.1109/MRA.2010.937861
8. Rashidi, P., Mihailidis, A.: A survey on ambient-assisted living tools for older adults. IEEE J. Biomed. Health Inform. **17**(3), 579–590 (2013). https://doi.org/10.1109/JBHI.2012.2234129
9. Swangnetr, M., Kaber, D.B.: Emotional state classification in patientrobot interaction using wavelet analysis and statistics-based feature selection. IEEE Trans. Hum. Mach. Syst. **43**(1), 63–75 (2013). https://doi.org/10.1109/TSMCA.2012.2210408

10. Zhang, Y., Qiu, M., Tsai, C., Hassan, M.M., Alamri, A.: Health-CPS: Healthcare cyber-physical system assisted by cloud and big data. IEEE Syst. J. **11**(1), 88–95 (2017). https://doi.org/10.1109/JSYST.2015.2460747
11. AMY-Robot. https://www.robotiko.it/amy-robot/
12. SERROGA project. https://www.tu-ilmenau.de/de/neurob/projects/finished-projects/serroga/

Selecting for Positive Responses to Knock Outs in Boolean Networks

Marco Villani[1,2(✉)], Salvatore Magrì[1], Andrea Roli[2,3], and Roberto Serra[1,2,4]

[1] Department of Physics, Informatics and Mathematics,
University of Modena and Reggio Emilia, Modena, Italy
marco.villani@unimore.it
[2] European Centre for Living Technology, Venice, Italy
[3] Department of Computer Science and Engineering, University of Bologna, Cesena, Italy
[4] Institute for Advanced Study, University of Amsterdam, Amsterdam, The Netherlands

Abstract. Random Boolean networks are a widely acknowledged model for cell dynamics. Previous studies have shown the possibility of achieving Boolean Networks (BN) with given characteristics by means of evolutionary techniques. In this work we show that it is possible to evolve BNs exhibiting more positive than negative reactions to knock-out stresses. It is also interesting to observe that in the observed runs (i) the evolutionary processes can guide the BNs toward different dynamic regimes, depending on their internal structure and that (ii) the BNs forced to evolve by maintaining a critical dynamical regime achieve better results than those that do not have this characteristic; this observation supports the idea that criticality may be beneficial to an evolving population of dynamical systems.

Keywords: Random Boolean networks · Gene knock outs · Evolved systems

1 Introduction

Random Boolean Networks (RBNs for short) are well-known models of gene regulatory dynamics [1–4]. At the same time, RBNs have a prominent role in the development of complexity science [5–9], being able to exhibit qualitatively different dynamical regimes (ordered, critical, disordered) depending on some structural parameters.

Let us consider the distribution of changes in gene expression levels induced by single-gene knock-outs, which has already been the subject of previous studies [1, 3, 10–12]. Deferring a wider discussion to Sect. 2, let us recall that a single-gene knock-out is the permanent silencing of a particular gene in a cell, so that the protein corresponding to that gene is not synthesized. A modification of this kind may affect other genes, which are said to be "affected" by that perturbation. The set of affected gene is the "avalanche" corresponding to that perturbation and its size is the ratio between the number of genes in the avalanche and the total number of genes.

In the past, we have shown that, notwithstanding their radical simplifying assumptions, RBNs are effective in simulating the actual distribution of avalanches in S. Cerevisiae [1, 3, 11, 12]. In those works, no distinction was made between the sign of the

© Springer Nature Switzerland AG 2020
F. Cicirelli et al. (Eds.): WIVACE 2019, CCIS 1200, pp. 7–16, 2020.
https://doi.org/10.1007/978-3-030-45016-8_2

induced changes, i.e. a gene was regarded as affected without distinguishing the cases where the expression level is increased (UP) from those where it is lowered (DOWN). However, when we look at the distribution of UPs and DOWNs in the biological systems, we observe that the formers are much more frequent than the latters – a property which is not easily reproduced in RBNs.

We therefore tried to evolve BNs, starting from RBNs, in such a way that the evolved networks have indeed this property. Previous studies have indeed shown the possibility of achieving Boolean Networks (BNs) with some desired characteristics by means of evolutionary techniques [13–18] starting from classical RBNs. The study described in [19] is particularly interesting here: it shows that it is possible to evolve networks with a desired fraction of active genes, which are critical during all the stages of the evolutionary process. This is important since critical networks have been suggested to have important advantages with respect to ordered and disordered ones, so it is likely that this property is conserved during the various stages of biological evolution. Moreover, in the same work a modified genetic algorithm was introduced, which maintains criticality, which is used also here for the same reasons of biological advantage. The algorithm is briefly summarized in Sect. 2.

We defer to the last section a discussion of the main results. Let it suffice here to mention that the goal has been attained, thus confirming that it is possible to evolve critical Boolean nets with non obvious properties. It has also been proven that the evolved networks are no longer random.

An intriguing question concerns the identification of the features of the evolved networks. We have considered some possible candidates, like the fraction of so-called canalizing functions and the frequency with which different types of Boolean rules are coupled. In this respect, further studies will be needed.

The paper is organized as follows: in Sect. 2 the methods are defined, while the experimental results are presented in Sect. 3 and discussed in Sect. 4.

2   Methods

Gene regulatory networks have been modelled by random Boolean networks (RBNs), which have been described and discussed at length in several papers and books [5–9, 20, 21]. Let it suffice here to recall that they are time-discrete dynamical models, where N boolean variables interact according to fixed rules (i.e., Boolean functions). The state at time $t + 1$ of the i-th node is determined by the states, at time t, of a fixed set of k input genes, and all the nodes are simultaneously updated. The set of input nodes is chosen at random with uniform probability, and the Boolean functions are also chosen at random with bias p: to each set of input values a value 0 or 1 is chosen with probability p.

RBNs can be either ordered, disordered or critical. A particular attention is due to critical networks since it has been suggested by several authors [5, 6, 8, 9, 22] that they can provide an optimal tradeoff between robustness and responsiveness to variations, so they should be selected by biological (and artificial?) evolution. Criticality is assessed here by the so-called Derrida parameter λ [23] which measures the average normalized Hamming distance at time $t + 1$ between two states which differ by the value of a single node at time t: a value $\lambda > 1$ (respectively, $\lambda < 1$) corresponds to disordered (respectively, ordered) networks, while a value $\lambda \approx 1$ is associated to critical networks.

It has been observed [24] that, in gene network models based upon actual biological observations, not all the different Boolean functions are represented, but that there is a strong bias in favor of the so-called canalizing functions. Following [25], a function is canalizing if there is at least one value of one of its inputs that determines the output. A function could be canalizing in more than one input, and the distribution of Boolean functions through the various canalizing classes has effects on the dynamic regime of the RBNs [26]. Canalizing functions for the case $k = 2$ are shown in Table 1.

Table 1. Scheme of Boolean functions and their classification in canalizing functions in one input (C1), in two inputs (C2), and in no input (C0)

INPUT		–	C1		C2				C0
Input A	Input B	FALSE	A	B	AND	A AND NOT B	NOT A AND B	OR	XOR
0	0	0	0	0	0	0	0	0	0
0	1	0	0	1	0	0	1	1	1
1	0	0	1	0	0	1	0	1	1
1	1	0	1	1	1	0	0	1	0
Input A	Input B	TRUE	NOT B	NOT A	A OR NOT B	NOT A OR B	NAND	NOR	NOT XOR
0	0	1	1	1	1	1	1	1	1
0	1	1	0	1	0	1	1	0	0
1	0	1	1	0	1	0	1	0	0
1	1	1	0	0	1	1	0	0	1

The availability of powerful experimental techniques (i.e. DNA microarrays) for the simultaneous measurement of the expression levels of thousands of genes makes it possible to analyze the patterns of gene activation in response to different environmental conditions, or to study – the theme of this work – the response of a system to the silencing of a single gene. Important sets of experiments are described in [27, 28] where many genes have been knocked-out (i.e. silenced), one at a time, in Saccharomyces Cerevisiae, and the changes in activations of all the known genes have been measured. In other works we already studied the distribution of the perturbation size and its relation with the dynamical regime of the underlying gene regulatory system (studied through the comparison with suitable RBN [1, 3, 11, 12, 29] or gene-protein systems [30, 31]).

It is worthwhile to remark that most affected genes (i.e. those whose expression level changes in response to the knock-out) in Saccharomyces Cerevisiae are of the UP type: indeed, almost 70% of the induced changes are in the direction of an increase in the activation of the genes. On the other hand, typical RBNs have a low probability of showing a similar response (as shown in Fig. 1). But yeasts have been subject to biological evolution for a long time, therefore it is interesting to verify whether it is possible to evolve the networks to achieve such a response. Moreover, it is particularly interesting to consider if this can be achieved while constraining the evolution to be limited to critical (or near critical) networks.

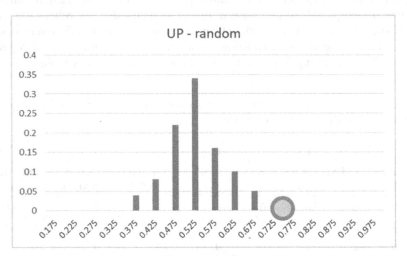

Fig. 1. The response to a single perturbation of 100 different RBNs. For each network, all possible knock-out were performed on the attractor with the largest attraction basin, and the fraction of genes (on the total disturbed genes) whose average activity grew (increase in activation - UP) in response to the changed situation is stored. The yellow dot shows the position of a real system (Saccharomyces Cerevisiae), value derived from the data present in [27, 28]

In order to induce an evolutionary process, we made use of the well-known Genetic Algorithms (GA), introduced by John Holland [32] (in the following we assume that the reader is familiar with these techniques – see also Table 2 for more details). The individuals are the single RBNs, the genotype being composed by the Boolean tables that guide the dynamic response of each node; in each GA run all RBNs share the same topology.[1] Each RBN is subject to a knock-out experiment (all possible knock-out of a single gene, performed on a random state of the RBN attractor owning the largest basin of attraction): the fitness is proportional to the fraction of genes involved in an avalanche showing an increase in activation.

The GA may be run without constraints ("free GA"), or may be constrained to maintain the bias of RBNs present in the first generation, the purpose of this last version being that of preserving the dynamic regime of the initial RBNs over the generations [19]. Since this result is not guaranteed, because the experiment involves evolved RBNs - and therefore no longer "random" BNs, for which there is a statistical link between the structural parameters and the dynamic regime - we verified the dynamic regime of each generated system over the generations. In all performed GA runs, the dynamical regime of the involved systems has indeed maintained its initial characteristic.

---

[1] Since the numbering of nodes is arbitrary (the number of a node does not carry any intrinsic meaning, like e.g. any reference to a specific biological gene), this does not limit the set of actually different networks that can be generated.

Table 2 The general scheme of the used GA (#G denotes the number of elements in set G). * In the case of balanced GA, the crossover point is chosen so as to keep the bias of the Boolean functions of children close to that of the parents [19]. ** In the case of balanced GA, mutations are carried out in order to maintain the bias of the Boolean functions of the children close to the parents' bias [19]

Step	Step description
1	Create network topology (it will be the same for all the networks for all generations)
2	Create the first population G of networks (Boolean functions are generated at random with bias p)
3	For each network in G, compute its fitness
4	Select the set E of the individuals with the highest fitness, which will be passed unaltered in the next generation (elitism)
5	Select parents for the individuals of the new generation, with probability proportional to their fitness (their number being equal to #G-#E)
6	Generate the set G' by applying single-point crossover* to the selected parents in G with a given probability (otherwise parents pass unmodified in the new population)
7	Generate the set G" by applying single-point mutation** to individuals in G' with a fixed (small) probability
8	Generate the new population of networks G = E ∪ G"
9	If termination condition has not been met, return to step 3
10	End

3   The Experiments

As described in Sect. 2 the GA could act without constraints ("free GA"), or could have the constraint of maintaining the bias of RBNs present in the first generation ("balanced GA"), preserving in such a way the RBN's initial dynamical regime [19]. Interestingly, the GA itself could act an evolutionary pressure on the process: in particular, the GAs we used acts symmetrically on the "1" and "0" of truth tables, creating or subtracting symbols in the same way. As consequence, even in case of the "free GA" RBNs with an average connectivity $k_{in} = 2$ and an initial bias of 0.5 typically maintain such a situation, whereas RBNs with an average connectivity $k_{in} = 3$ and an initial bias of 0.21 (both situations being "critical" in the Kauffman scheme [5, 6]) tends to increase the bias of their Boolean functions. In Fig. 2 we can observe this phenomenon, measured through the so called Derrida coefficient, and index assuming values lower than 1 for ordered systems and higher than 1 for disordered systems (values close to 1 indicating critical dynamical regimes) [23].

An even more interesting situation is shown in Fig. 3, which presents the overall behavior of the evolutions by using the classical GA and the balanced GA. In fact, while the two variants of the GA have no effect on the overall evolution of RBNs with average connectivity equal to 2, the same does not happen for RBNs with average connectivity equal to 3. Next to the already mentioned effect on the dynamic regime, the networks with

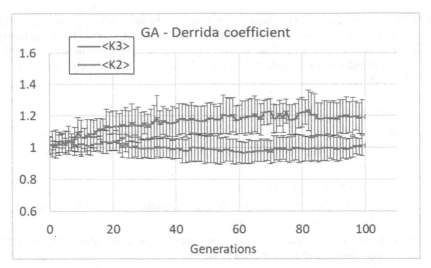

Fig. 2. GA progression: Derrida coefficient (averages on 10 different GA runs). Note that RBNs with <k> = 2 are kept in critical conditions, while RBNs with <k> = 3 are shifted because the evolutionary pressure toward more disordered dynamical regimes

k = 3 show a significant increase in fitness, both considering its average value between different runs (respectively 0.76 ± 0.04 and 0.90 ± 0.04 to the 250th generation), and considering its maximum within all runs (respectively 0.837 and 0.954 to the 250th generation). In this case the fact of maintaining a critical dynamic regime (by keeping the bias of the Boolean functions balanced) seems to provide a considerable support to the evolutionary capacity of the RBN population.

Eventually, evolution leads to (no longer random) BNs showing very high fractions of genes with positive response to the knock-outs. In this work we focus our attention to BNs having an input connectivity $k_{in} = 2$. Indeed, despite the fact they react to knock-outs in a way far from that typical of random networks, they do not show obvious structural characteristics different from random systems. In addition, even the categorization in canalization class [24], or the correlation between the Boolean functions directly linked through the systems' topology do not show particular issues (Fig. 4). In particular, the direct associations between the different types of Boolean functions (which type of Boolean function is present as inputs of which type of Boolean function) do not show significantly different trends from the completely random case (Table 3).

Nevertheless, the not evident – but present - organizational features make evolved networks quite peculiar. It is in fact possible to destroy the organization of evolved networks by creating variants in which i) the Boolean functions have been mixed (thus destroying the association between the different Boolean functions) or ii) new ones have been randomly created (thus destroying the association between Boolean functions and their distribution). These randomized systems have a fraction of genes exhibiting UP-responses to knock-outs very far from that of the evolved objects (Fig. 5).

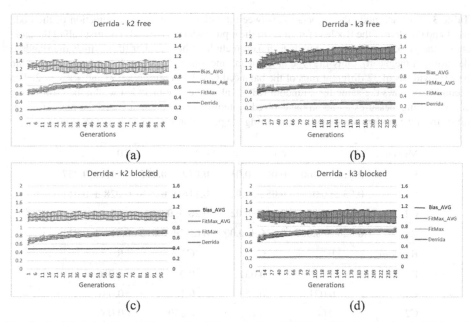

Fig. 3. A global overview of the results of the evolution of RBNs subject to the classical GA and the balanced GA: in each figure, from top to bottom, the Derrida coefficient, the maximum fitness (on 10 different GA runs), the average fitness with error bar, the bias of the Boolean functions. The secondary axis indicates the correct scale for the Derrida coefficient. First row, classical GA; second row, balanced GA. Left column, RBNs with average connectivity <k> = 2; right column: RBNs with average connectivity <k> = 3.

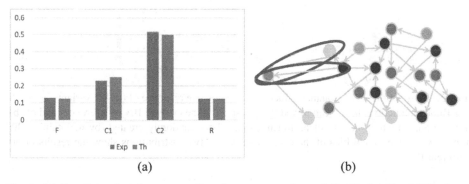

Fig. 4. (a) Canalizing functions: comparison between experimental distribution Exp (all Boolean functions of the best individuals of 32 different runs of the balanced GA, RBN with N = 50 and <k> = 2) and the theoretical distribution Th of random RBNs. (b) Analysis scheme used in Table 1: for each node the association between the Boolean function of the node itself and the Boolean functions of the input nodes is detected. Table 3 shows the probability associated with each pair (only the type of Boolean functions is considered).

Table 3. For each node the association between the class of the Boolean function of the node itself and the class of the Boolean functions of the input nodes is detected. The first half of the table shows the probability associated with each pair, including the error bar (three times the standard deviation of the mean). Only the type of Boolean functions is considered; the data are about the best individuals of 32 different runs of the balanced GA, RBN with N = 50 and <k> = 2. The second half shows the same for a random RBN – all values are included within the error of the corresponding measures made on the evolved systems. In bold are evidenced the measures that differ from the random situation by considering the plain standard deviation of the mean.

Measured	F	C1	C2	C0
F	0.015 ± 0.007	0.03 ± 0.01	**0.070 ± 0.015**	**0.013 ± 0.007**
C1	0.03 ± 0.01	**0.05 ± 0.02**	**0.118 ± 0.018**	0.028 ± 0.011
C2	**0.07 ± 0.02**	**0.11 ± 0.02**	0.26 ± 0.04	0.07 ± 0.02
R	**0.020 ± 0.007**	**0.026 ± 0.009**	0.064 ± 0.015	0.014 ± 0.011

Random	F	C1	C2	R
F	0.016	0.031	0.063	0.016
C1	0.031	0.063	0.125	0.031
C2	0.063	0.125	0.250	0.063
R	0.016	0.031	0.063	0.016

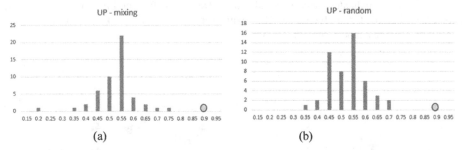

Fig. 5. Distribution of the avalanche fractions with positive response of (a) BN variants in which the Boolean functions have been mixed (with respect to the evolved BN of reference) and of (b) BN variants in which new Boolean functions have been randomly created (by maintaining the topology and the global bias of the evolved BN). The two distributions show the results of 50 different BNs

4   Conclusions

It is has been shown that it is possible to evolve BNs to have the characteristic of showing more positive than negative reactions to knock-out stresses. The networks which are obtained are no longer truly random, although they still show some degree of randomness. It has been shown that the evolved networks cannot be characterized by the fraction of canalizing functions, nor by the two-point correlations between types of successive

Boolean functions. Similar observations have been made in the past in the case of networks evolved to show a given density of "1"s in their attractors. Further work is needed to uncover the "secret" of networks endowed with these properties; this might lead to very interesting findings concerning the important features of Boolean networks.

It is also interesting to observe that, in the case $k = 3$, networks evolved according to the balanced GA achieve better results than those evolved by the free GA; this lends support to the idea that criticality may be beneficial to an evolving population of dynamical systems.

References

1. Serra, R., Villani, M.: Semeria A Genetic network models and statistical properties of gene expression data in knock-out experiments. J. Theor. Biol. **227**, 149–157 (2004)
2. Shmulevich, I., Kauffman, S.A.: Aldana M Eukaryotic cells are dynamically ordered or critical but not chaotic. PNAS **102**(38), 13439–13444 (2005)
3. Serra, R., Villani, M., Graudenzi, A.: Kauffman SA Why a simple model of genetic regulatory networks describes the distribution of avalanches in gene expression data. J. Theor. Biol. **246**(3), 449–460 (2007)
4. Villani, M., Barbieri, A.: Serra R A dynamical model of genetic networks for cell differentiation. PLoS ONE **6**(3), e17703 (2011)
5. Kauffman, S.A.: The Origins of Order. Oxford University Press, Oxford (1993)
6. Kauffman, S.A.: At Home in the Universe. Oxford University Press, Oxford (1995)
7. Bastolla, U., Parisi, G.: The modular structure of Kauffman networks. Phys. D **115**(3–4), 219–233 (1998)
8. Bastolla, U., Parisi, G.: Relevant elements, magnetization and dynamical properties in Kauffman networks: a numerical study. Phys. D **115**(3–4), 203–218 (1998)
9. Aldana, M., Coppersmith, S., Kadanoff, L.P.: Boolean dynamics with random couplings. In: Kaplan, E., Marsden, J., Sreenivasan, K.R. (eds.) Perspectives and Problems in Nonlinear Science. Springer, New York (2003). https://doi.org/10.1007/978-0-387-21789-5_2
10. Serra, R., Villani, M., Graudenzi, A., Colacci, A., Kauffman, S.A.: The simulation of gene knock-out in scale-free random Boolean models of genetic networks. Netw. Heterogen. Media **3**(2), 333–343 (2008)
11. Di Stefano, M.L., Villani, M., La Rocca, L., Kauffman, S.A., Serra, R.: Dynamically critical systems and power-law distributions: Avalanches revisited. In: Rossi, F., Mavelli, F., Stano, P., Caivano, D. (eds.) WIVACE 2015. CCIS, vol. 587, pp. 29–39. Springer, Cham (2016). https://doi.org/10.1007/978-3-319-32695-5_3
12. Villani, M., La Rocca, L., Kauffman, S.A., Serra, R.: Dynamical criticality in gene regulatory networks. Complexity **2018**, 14 p. (2018). Article ID 5980636
13. Liu, M., Bassler, K.E.: Emergent criticality from coevolution in random Boolean networks. Phys. Rev. E Stat. Nonlin. Soft. Matter Phys. **74**, 041910 (2006)
14. Szejka, A., Drossel, B.: Evolution of canalizing Boolean networks Eur. Phys. J. B **56**, 373–380 (2007)
15. Mihaljev, T., Drossel, B.: Evolution of a population of random Boolean networks Eur. Phys. J. B **67**, 259 (2009)
16. Gershenson, C.: Guiding the self-organization of random Boolean networks. Theory Biosci. **131**, 181–191 (2012)
17. Benedettini, S., et al.: Dynamical regimes and learning properties of evolved Boolean networks. Neurocomputing **99**, 111–123 (2013)

18. Braccini, M., Roli, A., Villani, M., Serra, R.: Automatic design of boolean networks for cell differentiation. In: Rossi, F., Piotto, S., Concilio, S. (eds.) WIVACE 2016. CCIS, vol. 708, pp. 91–102. Springer, Cham (2017). https://doi.org/10.1007/978-3-319-57711-1_8

19. Magrì, S., Villani, M., Roli, A., Serra, R.: Evolving critical boolean networks. In: Cagnoni, S., Mordonini, M., Pecori, R., Roli, A., Villani, M. (eds.) WIVACE 2018. CCIS, vol. 900, pp. 17–29. Springer, Cham (2019). https://doi.org/10.1007/978-3-030-21733-4_2

20. Kauffman, S.A.: Metabolic stability and epigenesis in randomly constructed genetic nets. J. Theor. Biol. **22**, 437–467 (1969)

21. Drossel, B.: Random Boolean networks. In: Schuster, H.G. (ed.) Reviews of Nonlinear Dynamics and Complexity, vol. 1, pp. 69–110. Wiley, Weinheim (2008)

22. Aldana, M., Balleza, E., Kauffman, S.A., Resendiz, O.: Robustness and evolvability in genetic regulatory networks. J. Theor. Biol. **245**(3), 433–448 (2007)

23. Derrida, B., Pomeau, Y.: Random networks of automata: A simple annealed approximation. Europhys. Lett. **1**(2), 45–49 (1986)

24. Harris, S.E., Sawhill, B.K., Wuensche, A., Kauffman, S.A.: A model of transcriptional regulatory networks based on biases in the observed regulation rules. Complexity **7**, 23–40 (2002)

25. Just, W., Shmulevich, I., Konvalina, J.: The number and probability of canalizing functions. Phys. D **197**, 211–221 (2004)

26. Karlsson, F., Hornquist, M.: Order and chaos in Boolean gene networks depends on the mean fraction of canalizing functions. Phys. A **384**, 747–755 (2007)

27. Hughes, T.R., Marton, M.J., Jones, A.R., et al.: Functional discovery via a compendium of expression profiles. Cell **102**(1), 109–126 (2000)

28. Kemmeren, P., Sameith, K., van de Pasch, L.A.L., et al.: Largescale genetic perturbations reveal regulatory networks and an abundance of gene-specific repressors. Cell **157**(3), 740–752 (2014)

29. Roli, A., Villani, M., Filisetti, A., Serra, R.: Dynamical criticality: Overview and open questions. J. Syst. Sci. Complex **31**(3), 647–663 (2018)

30. Graudenzi, A., Serra, R., Villani, M., Colacci, C., Kauffman, S.A.: Robustness analysis of a Boolean model of gene regulatory network with memory. J. Comput. Biol. **18**(4), 559–577 (2011). Mary Ann Liebert, Inc., Publishers, NY

31. Sapienza, D., Villani, M., Serra, R.: Dynamical properties of a gene-protein model. In: Pelillo, M., Poli, I., Roli, A., Serra, R., Slanzi, D., Villani, M. (eds.) WIVACE 2017. CCIS, vol. 830, pp. 142–152. Springer, Cham (2018). https://doi.org/10.1007/978-3-319-78658-2_11

32. Holland, J.H.: Adaptation in Natural and Artificial Systems. University of Michigan Press, Ann Arbor (1975)

Avalanches of Perturbations in Modular Gene Regulatory Networks

Alberto Vezzani[1], Marco Villani[1,2(✉)], and Roberto Serra[1,2,3]

[1] Department of Physics, Informatics and Mathematics, Modena and Reggio Emilia University,
Modena, Italy
marco.villani@unimore.it
[2] European Centre for Living Technology, Venice, Italy
[3] Institute for Advanced Study, University of Amsterdam, Amsterdam, The Netherlands

Abstract. A well-known hypothesis, with far-reaching implications, is that biological evolution should preferentially lead to critical dynamic regimes. Useful information about the dynamical regime of gene regulatory networks can be obtained by studying their responses to small perturbations. The interpretation of these data requires the use of suitable models, where it is usually assumed that the system is homogeneous. On the other hand, it is widely acknowledged that biological networks display some degree of modularity, so it is interesting to ascertain how modularity can affect the estimation of their dynamical properties. In this study we introduce a well-defined degree of modularity and we study how it influences the network dynamics. In particular, we show how the estimate of the Derrida parameter from "avalanche" data may be affected by strong modularity.

Keywords: Gene regulatory networks · Random Boolean Networks · Dynamical regimes · Modular networks

1 Introduction

The idea that critical dynamic regimes possess some peculiar advantages, so that they tend to be selected under evolutionary dynamics, has relevant implications in biology [1, 2] as well as in evolutionary computation and in the design of artificial systems [3]. This idea, often referred to as the criticality hypothesis, has been suggested by several authors [1–7] and has played a noticeable role in the search for general principles in complexity science [8, 9].

There are many characteristics related to dynamic regimes: in this work we refer to the response of the asymptotic state of the system to a small perturbation, which can be classified as ordered, if the perturbation dies out in time, or disordered if it increases for some finite time interval. A system is said to be critical if its behaviour is intermediate between order and disorder. Note that all the above statements refer to average behaviours [1, 2, 10, 11].

In recent times the great availability of biological data and the use of models has made it possible to deal with the problem of testing the criticality hypothesis.

© Springer Nature Switzerland AG 2020
F. Cicirelli et al. (Eds.): WIVACE 2019, CCIS 1200, pp. 17–27, 2020.
https://doi.org/10.1007/978-3-030-45016-8_3

In particular, part of the research takes advantage of the data coming from molecular biology (mainly data concerning genetic regulatory networks, or GRN [12]) and of a well-known model (that of the Random Boolean Network, in the following RBN [1, 2, 10, 13, 14]), one of the most used frameworks in the complex systems research field.

Valuable information regarding the dynamical regime of a GRN can be obtained by the study of the distribution of avalanches of perturbations of gene expression levels, which follow the externally forced knock-out of a single gene [15, 16]. Knocking-out (i.e. permanent silencing) a single gene in a cell may affect the expression levels also of other genes: the set of affected genes is called the "avalanche" associated to that specific knock-out, and the number of affected genes is the size of the avalanche.

In a series of papers [17–21] we have shown that dynamical networks of Boolean nodes are able to describe actual (suitably booleanized) experimental avalanches of perturbations in the yeast Saccharomyces Cerevisiae [15, 16]. Moreover, it has been shown that, under reasonable assumptions, it is possible to use those data to draw inferences about the value of the so-called Derrida parameter (see Sect. 2), which describes the dynamical regime of the system [22, 23]. The "reasonable assumptions" which are required are that the genetic network is sparse (i.e. the number of links per node is much smaller than the number of nodes) and that there is no self-interference, i.e. that each node is reached by the perturbation only once (for a precise definition, see [21]). Both conditions seem to be satisfied in the case of S. Cerevisiae.

There is however a further hypothesis which has been implicitly assumed, i.e. that the overall GRN has uniform properties: this hypothesis is necessary in order to relate the avalanche distribution to the Derrida parameter. In this paper, we study a case where uniformity cannot be assumed, i.e. that of a modular network.

Indeed, it is widely believed that biological networks are modular [24–26], although the identification of the proper modules is a difficult and controversial task, given that modules are not completely independent, but do interfere with each other.

In the case of modular networks, the relationship between the overall Derrida parameter and the avalanche of perturbations no longer holds: indeed, the former is based on local perturbations, which tend to remain mostly confined within the module of the perturbed node, while the latter can spread further if the coupling strength between modules is high enough. However, in Sect. 3 it will be shown that the intra-modules redirection of a small fraction of nodes suffices to make the estimation error of the Derrida parameter from the size distribution of the permanent perturbations quite small, providing a value quite close to the one obtained by the classical procedure, based upon the study of the time evolution of a small transient perturbation.

This observation, as discussed in conclusions, may also suggest that by looking at the distribution of avalanches one can infer some information about the presence and the size of some relevant modules (although the experimental data may be blurred and difficult to use). In any case, comparing the dynamical response to local perturbations to system-wide perturbations like avalanches may be a royal road to uncover interesting information hidden in the data.

In Sect. 2 we provide the necessary details about the RBN framework, the criticality measure we use (the Derrida coefficient), and the modular topology. In Sect. 3 we present and discuss the results obtained working with modular systems; Sect. 4 contains the conclusions of the work.

2   Random Boolean Networks

Random Boolean Networks (RBNs for short) are a widely acknowledged model for gene regulatory dynamics [17, 18, 21, 27–29]. At the same time, RBNs have a prominent role in the development of complexity science, being able to exhibit very different dynamical regimes (ordered, critical, disordered) depending on some structural parameters.

In this section we present a synthetic description of the RBN model we are using, referring the reader to [1, 2, 10, 13, 14] for a more detailed account. Several variants of the model have been presented and discussed [30–34], but we will restrict our attention here to the "classical" model. A classical RBN is a dynamical system composed of N genes, or nodes, which can take either the value 0 (inactive) or 1 (active). Let $x_i(t) \in \{0, 1\}$ be the activation value of node i at time t, and let $X(t) = [x_1(t), x_2(t), \dots x_N(t)]$ be the vector of activation values of all the genes.

The relationships between genes are represented by directed links; the dynamic reaction to the stimuli of the upstream nodes are modelled through Boolean functions. In a classical RBN each node has the same number of incoming connections kin, whose sources chosen at random with uniform probability among the remaining $N - 1$ nodes. The resulting distribution of the outgoing connections per node tends therefore to a Poisson distribution for large N.

The Boolean functions can be chosen in different ways. In this work, for each node i and each of the 2^{kin} possible combinations of values of input nodes, the output value is assigned at random, choosing 0 with probability p and 1 with probability $1 - p$. The parameter p is called bias.

In the so-called quenched model, both the topology and the Boolean function associated to each node do not change in time. The network dynamics are discrete and synchronous, so fixed points and cycles are the only possible asymptotic states in finite networks. A single RBN can have, and usually has, more than one attractor. The model shows two main dynamical regimes, ordered and disordered, depending upon the degree of connectivity and upon the Boolean functions. Typically, the average cycle length grows as a power of the number of nodes N in the ordered region and diverges exponentially in the disordered region. The dynamically disordered region also shows sensitive dependence upon the initial conditions, not observed in the ordered one [1, 2].

It should be mentioned that some interesting analytical results have been obtained by the annealed approach, in which the topology and the Boolean functions associated to the nodes change at each step. Although the annealed approximation may be useful for analytical investigations [10], in this work we will always be concerned with quenched RBNs, which are closer to real gene regulatory networks.

In order to quantitatively determining the dynamic regime of an RBN, the so-called Derrida parameter λ can be used. This parameter can be determined by studying the average behaviour in time of the distance between two close initial states of a RBN and taking its limiting value when the initial distance approaches its limiting lower value (dynamical sensitivity) [22, 23]. In practice, one often considers the short time evolution of the distance between two initial states that differ by a single spin flip. On average, after one time step in "ordered" networks this distance is lower than 1 (a situation corresponding to a Derrida parameter $\lambda < 1$) and the small perturbation tends to vanish, while in chaotic networks ($\lambda > 1$) it increases in time, and in critical networks ($\lambda = 1$) it tends to maintain its size.

An alternative way to estimate the λ parameter is through the distribution of the size of the perturbation that is possible permanently induce within a RBN. For a given permanent silencing, let us define a node as "affected" if its value is significantly modified with respect to the wild type. Let then an avalanche be the set of affected nodes, and define the size of the avalanche as the number of the nodes that are affected. The permanent silencing models the so-called "gene knock-out" within a real living system: and it is possible to demonstrate that, under suitable assumptions, the size distribution of the avalanches made on a RBN depends only upon the RBN Derrida parameter through the formula:

$$p(v) = \frac{v^{v-2}}{(v-1)!}\lambda^{v-1}e^{-\lambda v} \tag{1}$$

where $p(v)$ is the probability that a randomly chosen avalanche's size is v, and λ is the Derrida parameter [19]. This consideration has been discussed in a series of papers in the case of RBN simulating the dynamic regime of the yeast S. Cerevisiae [17–21] - the best estimate of the Derrida parameter places it in the ordered region, not far from the critical boundary (for more details see [21]).

If the network is critical (i.e. $\lambda = 1$) then, by adopting the well-known Stirling formula, the above formula approximates a power-law in the case of not-too-small avalanches

$$p(v) = \frac{1}{\sqrt{2\pi}}v^{-\frac{3}{2}} \tag{2}$$

Note that in this way it is possible to use a static measure (the size distribution of avalanches) to draw inferences about the dynamical regime of the network. Note also the nature of the avalanches, which - unlike the Derrida methodology - show the effects of a perturbation after a significant amount of time.

As just discussed, "classic" RBNs have a random homogeneous topology. However, the dynamic response of a system can be strongly influenced by its topology, and there are indications that in biology different topologies may be more frequent than the homogeneous one [24–26, 35, 36].

In this paper we focus on the dynamic effects of the introduction of a topology in which modules are present (a module loosely described as an identifiable part of a network weakly connected to the rest of the system).

So, while uniformity may be a useful initial ("null") hypothesis, the exploration of the effects of modularity is very interesting. In this paper, we introduce a modular network that is perfectly known, and we analyse the distribution of avalanches while varying the coupling between the modules. The basic intuition is that, when the modules are almost independent, the perturbations spreads only in the module where the knock-out takes place, and they very rarely affect nodes in other modules. As the interactions between different modules grow, one observes that the fraction of avalanches that affect the other modules increases. By increasing the number of links between different modules we can tune the network from fully disconnected (composed by independent modules) to fully homogeneous. It is then interesting to ask how the estimation of the properties of interest is affected by the degree of inter-module connections. In particular, we will consider here the way in which it influences the estimation of the Derrida parameter from avalanche size distributions.

In this work the procedure used to build a modular network is composed by the following steps:

1. **Generation of N_m (number of modules) RBN** with a constant k_{in} (in the following experiments all the nodes have the same input connectivity) and bias. In particular, in this paper we consider the "classical" critical parameters of $k_{in}= 2$ and bias = 0.5. The behaviour of networks with different N_m number of modules has been considered: in this work however we show the case of networks when $N_m = 4$, the qualitative behaviour of different configurations being similar.
2. **Modules connection** through the redirection of a subset of their links. For every module, this process randomly select *n-redirect* input links and change their source to a random node of a different module.[1] In addition, for each connection from module A to module B we also create a connection from module B to module A (using two different nodes from the previous ones), in order to maintain a degree of symmetry between the modules. The identifiers of the n-redirect links are stored in a suitable matrix[2].

The topology of the connections between modules could considerably affect the behaviour of the whole system. In this paper we consider as first step of a more complex analysis a random topology, where each module (with the obvious exception of the target module) has the same probability of being present as a source of one of the *n-redirect* links, by maintaining in such a way the same connection strategy used for the single modules.

3 Results

In the following we focus our attention to modular networks with 200 nodes[3], bias = 0.5, $k_{in} = 2$ (critical), where internal modules – each module being a complete RBN - are connected each other through a variable number of links (*n-redirect* links). Several different N_m number of modules has been considered: in this paper we show the case of networks when $N_m = 4$ (therefore composed by 50 nodes each), the qualitative behaviour

---

[1] Note that in such a way the activation function of every node is kept unchanged, the only alteration being the source of the input link. In this way the process preserves the contribution of the nodes to the overall dynamic regime.

[2] All links are chosen in order to prevent multiple redirections.

[3] In any case, the behavior of networks with different number of nodes has been considered without noting qualitatively different behaviors.

of the other configurations (2 modules of 100 nodes each, 2 modules of 67 nodes each and 1 module of 66 nodes, and 8 modules of 25 nodes each) being similar.

The topologies studied are obtained by redirecting *n-redirect* input links to exist between different modules, *n-redirect* belonging to the set (when possible): {0, - separated modules-, 1, 2, 3, 5, 10, 15, 20, 25}. The initial case of independent modules is then compared with the cases where a number of links has been redirected. The number of redirections grows up to a case where each node receives in average a quarter of their incoming inputs from the module to which it belongs, and three quarters from the other modules

The experiments were done as follows:

1. **Derrida's coefficient**: for each topology, this study considers the evolution of 100 networks, each network starting from 10000 initial conditions randomly generated (0.5 chance for every node to be activated). For each initial condition we performed 100 measures of Hamming distance between perturbed and unperturbed state (we perturbed 100 times a different node), so the contribution to the Derrida parameter value of each initial condition comes from an average of 100 numbers. The final Derrida parameter is the average of these measures over the 10000 initial conditions. Figure 1b shows the results for the random connection of 4 modules of 50 nodes each, when they are randomly connected by {0, - separated modules-, 1, 2, 3, 4, 5, 10, 15, 20, 25} couples of links.

2. **Avalanches**: for each topology, knock-outs are made on 20000 networks generated just in time, with the same specifics. Every launch evolves each single network to it's attractor, creates a copy of the system, then clamps the copy an active node to inactive and evolves again both the structures. Data about the number of nodes who changed their average activity (avalanche size) between the two new trajectories is stored and analysed.

In order to correctly interpret the measures performed on the analysed systems we need a comparison case: in this paper we use a random and homogeneous (without modules) system composed of 200 nodes (bias $= 0.5$, $k_{in} = 2$). This system is analysed by using the same procedure of the modular systems.[4] As expected, the dynamical behaviour of this system is close to criticality (Fig. 1a shows that the Derrida parameter of system "200_control" is close to 1.0); interestingly, the same analysis reports the value 1.0 for all modular systems, irrespectively to their internal organisation (Fig. 1a).

Low connectivity between modules strongly limits the presence and the size of large avalanches (Fig. 2). This fact in turn affect the estimate of the probabilities of occurrence of small size avalanches (because of the usual normalisation, so that the final distribution area is equal to 1).

An evident feature is the absence of large avalanches: the lower is the coupling, the poorer is the fraction of large avalanches – a nonlinear phenomenon that shows saturation effects. Note however that a small fraction of rewirings suffices to make the value of the

---

[4] That is, 100 different networks, each network measured in 10000 initial conditions for the Derrida parameter estimate, and 20000 different networks in order to obtain the avalanche distribution.

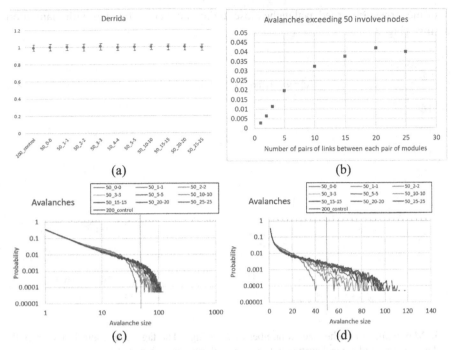

Fig. 1. The results of the analyses of the examined systems (the control system "200_control", and the modular systems with 0, 1, 2, 3, 5, 10, 15, 20, 25 couples of links) (a) Derrida parameter. (b) Fraction of avalanches whose size is higher than 50 (the extent of the used modules), depending on the number of pairs of links between each pair of modules (c) Avalanches distribution (we can observe the central "body" following a power law - a straight-line in a log-log plot). (d) The same avalanches distribution, linear-logarithmic plot

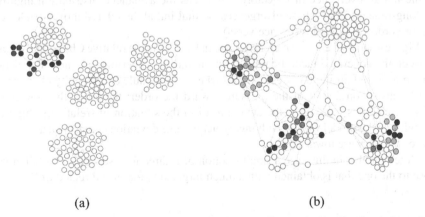

Fig. 2. Modular systems and spread of perturbations. (a) A system composed by four independent modules: an avalanche of changes is confined on the module where the initial perturbation is performed. (b) A system composed by four interdependent modules: the avalanches of changes have the possibility of spreading in different modules. The dark dots indicate the nodes that are decreasing their activity in response to the initial perturbation or to the subsequent activity changes of other nodes; the red dots indicate the nodes that are increasing their activity; the white nodes are not affected by the perturbation. (Color figure online)

size of the largest observed avalanche close to the one that is obtained with many more rewirings (Fig. 3).

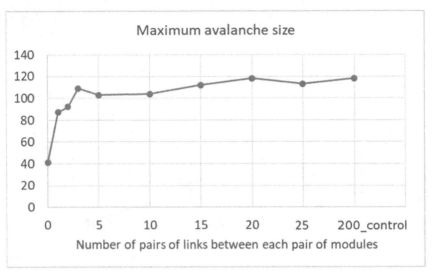

Fig. 3. Maximum avalanche size vs number of rewirings. The tag "200_control" indicates the completely connected control system, where a modular organisation is absent

The modification of the distribution of avalanches induced by the modularity however affects the estimates of the dynamical regime based upon it: this phenomenon highlights a significant issue, that is, the different kind of dynamical analysis performed by the two systems under consideration in this article. The Derrida methodology in fact is focusing on the initial divergence of trajectories, whereas the avalanche distribution highlights the long-time effects of these divergences, so that initial disordered moves could result in more ordered dynamics (or vice versa).

Figure 4 shows the λ parameter estimate in Eq. 1 (and the relative Chi Square distance between the generated theoretical distribution and the experimental one – the distance to be minimized during the λ estimation) when intermodule connectivity changes. The effect consists on a shift of the measure toward the order: the lower the connectivity between modules, the higher the deviation from the situation of (relatively) high connectivity. In the model we are studying, typically, this deviation consists on a systematic bias of the measure toward order.

Note that also in this case a small fraction of redirections suffices to obtain a value close to the one that is obtained with a much larger fraction of redirected links.

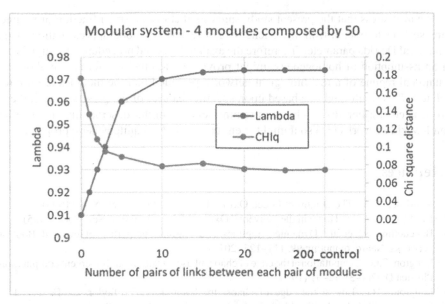

Fig. 4. Estimated Derrida coefficient minimizing (gradient descent, based on the Chi Square distance between the theoretical distribution depending on the λ parameter and the experimental one) the value of the distance from the experimental distribution. Starting from about a third of the links coming from other modules ($50*2 = 100$ incoming links to the nodes of a module, $10*(4 - 1) = 30$ links coming from other modules) the measurements saturate ("200_control" is the completely random network, where a modular organisation is absent)

4 Conclusions

In this paper we have shown the different behaviour of two quantities, which are both related to the dynamical regime of a Boolean model, in the case of modular networks. The Derrida parameter describes the short time response of the system under a small local perturbation, and is therefore largely insensitive to the modular structure. On the contrary, avalanches are long-term responses, and they can (or cannot) spread through the network, under appropriate conditions. Therefore the avalanche distribution can be heavily affected by the large-scale features of the network, and in particular by its modular organization.

We have shown these results working on an artificial model, where modules are hard-wired from outside. In the case of actual biological systems, we suggest that looking at their avalanche distribution might provide clues to their modular organization. For example, the distribution of small avalanches might approximately follow the same distribution of homogeneous networks, since they are smaller than the size of the modules, while the larger avalanches might be limited by the dimension of the modules. Such comparisons might then provide suggestions about the size of the modules. These are just preliminary hints, but they might be helpful in suggesting a way to use laboratory data to infer network properties.

We also stress that the present study suggests that even a small fraction of redirections suffices to obtain, from avalanche data, a reasonably good approximation of the dynamical Derrida parameter. Therefore the avalanche-based procedure, although based on an assumption of homogeneity, might provide quite precise information about the dynamical regime of a real biological network, provided that the modules are not too isolated. This result might be based upon specific features of the present model, therefore it needs further study – but it is reminiscent of similar observations in so-called small-world networks [37] so it might even turn out to be a quite generic property.

References

1. Kauffman, S.A.: The Origins of Order. Oxford University Press, New York (1993)
2. Kauffman, S.A.: At Home in the Universe. Oxford University Press, New York (1995)
3. Benedettini, S., et al.: Dynamical regimes and learning properties of evolved Boolean networks. Neurocomputing **99**, 111–123 (2013)
4. Langton, C.G.: Computation at the edge of chaos: phase transitions and emergent computation. Physica D **42**(1–3), 12–37 (1990)
5. Langton, C.G.: Life at the edge of chaos. In: Langton, C.G., Taylor, C., Farmer, J.D., Rasmussen, S. (eds.) Artificial Life II, pp. 41–91. Addison-Wesley, Reading (1992)
6. Packard, N.H.: Adaptation toward the edge of chaos. In: Dynamic Patterns in Complex Systems, pp. 293–301. World Scientific (1988)
7. Shmulevich, I., Kauffman, S.A., Aldana, M.: Eukaryotic cells are dynamically ordered or critical but not chaotic. PNAS **102**(38), 13439–13444 (2005)
8. Bar-Yam, Y.: Dynamics of Complex Systems. Addison-Wesley, Reading (1997)
9. Nicolis, G., Nicolis, C.: Foundations of Complex Systems: Nonlinear Dynamics, Statistical Physics, Information and Prediction. World Scientific, Singapore (2007)
10. Aldana, M., Coppersmith, S., Kadanoff, L.P.: Boolean dynamics with random couplings. In: Kaplan, E., Marsden, J.E., Sreenivasan, K.R. (eds.) Perspectives and Problems in Nonlinear Science, pp. 23–89. Springer, Heidelberg (2003). https://doi.org/10.1007/978-0-387-217 89-5_2
11. Kaneko, K.: Life: An Introduction to Complex Systems Biology. Springer, New York (2006)
12. Babu, M.M., Luscombe, N.M., Aravind, L., Gerstein, M., Teichmann, S.A.: Structure and evolution of transcriptional regulatory networks. Curr. Opin. Struct. Biol. **14**(3), 283–291 (2004)
13. Bastolla, U., Parisi, G.: The modular structure of Kauffman networks. Physica D **115**(3–4), 219–233 (1998a)
14. Bastolla, U., Parisi, G.: Relevant elements, magnetization and dynamical properties in Kauffman (1998b)
15. Hughes, T.R., Marton, M.J., Jones, A.R., et al.: Functional discovery via a compendium of expression profiles. Cell **102**(1), 109–126 (2000)
16. Kemmeren, P., Sameith, K., van de Pasch, L.A.L., et al.: Largescale genetic perturbations reveal regulatory networks and an abundance of gene-specific repressors. Cell **157**(3), 740–752 (2014)
17. Serra, R., Villani, M.: Semeria A Genetic network models and statistical properties of gene expression data in knock-out experiments. J. Theor. Biol. **227**, 149–157 (2004)
18. Serra, R., Villani, M., Graudenzi, A., Kauffman, S.A.: Why a simple model of genetic regulatory networks describes the distribution of avalanches in gene expression data. J. Theor. Biol. **246**(3), 449–460 (2007)

19. Serra, R., Villani, M., Graudenzi, A., Colacci, A., Kauffman, S.A.: The simulation of gene knock-out in scale-free random boolean models of genetic networks. Netw. Heterogen. Media **3**(2), 333–343 (2008)
20. Di Stefano, M.L., Villani, M., La Rocca, L., Kauffman, S.A., Serra, R.: Dynamically critical systems and power-law distributions: avalanches revisited. In: Rossi, F., Mavelli, F., Stano, P., Caivano, D. (eds.) WIVACE 2015. CCIS, vol. 587, pp. 29–39. Springer, Cham (2016). https://doi.org/10.1007/978-3-319-32695-5_3
21. Villani, M., La Rocca, L., Kauffman, S.A., Serra, R.: Dynamical criticality in gene regulatory networks. Complexity **2018**, 14 pages, Article ID 5980636 (2018)
22. Derrida, B., Pomeau, Y.: Random networks of automata: a simple annealed approximation. Europhys. Lett. **1**(2), 45–49 (1986)
23. Derrida, B., Flyvbjerg, H.: The random map model: a disordered model with deterministic dynamics. J. Phys. **48**(6), 971–978 (1987)
24. Ravasz, E., Somera, A.L., Mongru, D.A., Oltvai, Z.N., Barabasi, A.L.: Hierarchical organization of modularity in metabolic networks. Science **297**, 1551–1555 (2002)
25. Shen-Orr, S.S., Milo, R., Mangan, S., Alon, U.: Network motifs in the transcriptional regulation network of Escherichia coli. Nat. Genet. **31**, 64–68 (2002)
26. Gyorgy, A., Del Vecchio, D.: Modular composition of gene transcription networks. PLoS Comput. Biol. **10**(3), e1003486 (2014)
27. Damiani, C., Kauffman, S.A., Serra, R., Villani, M., Colacci, A.: Information transfer among coupled random boolean networks. In: Bandini, S., Manzoni, S., Umeo, H., Vizzari, G. (eds.) ACRI 2010. LNCS, vol. 6350, pp. 1–11. Springer, Heidelberg (2010). https://doi.org/10.1007/978-3-642-15979-4_1
28. Serra, R., Villani, M., Barbieri, B., Kauffman, S.A., Colacci, A.: On the dynamics of random boolean networks subject to noise: attractors, ergodic sets and cell types. J. Theoret. Biol. **265**, 185–193 (2010)
29. Villani, M, Barbieri, A, Serra, R.: A dynamical model of genetic networks for cell differentiation. PLoS ONE **6**(3), e17703 (2011). https://doi.org/10.1371/journal.pone.0017703
30. Glass, L., Kauffman, S.A.: The logical analysis of continuous, non-linear biochemical control networks. J. Theoret. Biol. **39**(1), 103–129 (1973)
31. Serra, R., Villani, M., Salvemini, A.: Continuous genetic networks. Parallel Comput. **27**, 663–683 (2001)
32. Graudenzi, A., Serra, R., Villani, M., Damiani, C., Colacci, A., Kauffman, S.A.: Dynamical properties of a Boolean model of gene regulatory network with memory. J. Comput. Biol. **18**, 1291–1303 (2011)
33. Graudenzi, A., Serra, R., Villani, M., Colacci, A., Kauffman, S.A.: Robustness analysis of a Boolean model of gene regulatory network with memory. J. Comput. Biol. **18**(4), 559–577 (2011)
34. Sapienza, D., Villani, M., Serra, R.: Dynamical properties of a gene-protein model. In: Pelillo, M., Poli, I., Roli, A., Serra, R., Slanzi, D., Villani, M. (eds.) WIVACE 2017. CCIS, vol. 830, pp. 142–152. Springer, Cham (2018). https://doi.org/10.1007/978-3-319-78658-2_11
35. Guido, N.J., Wang, X., Adalsteinsson, D., McMillen, D., Hasty, J., et al.: A bottom-up approach to gene regulation. Nature **439**, 856–860 (2006)
36. Purnick, P.E.M.: Weiss R The second wave of synthetic biology: from modules to systems. Nat. Rev. Mol. Cell Biol. **10**, 410–422 (2009)
37. Watts, D.J., Strogatz, S.H.: Collective dynamics of 'small-world' networks. Nature **393**, 440–442 (1998)

The Effects of a Simplified Model of Chromatin Dynamics on Attractors Robustness in Random Boolean Networks with Self-loops: An Experimental Study

Michele Braccini[1]([✉])(iD), Andrea Roli[1,4](iD), Marco Villani[2,4](iD), Sara Montagna[1](iD), and Roberto Serra[2,3,4](iD)

[1] Department of Computer Science and Engineering,
Alma Mater Studiorum Università di Bologna, Cesena, Italy
m.braccini@unibo.it
[2] Department of Physics, Informatics and Mathematics,
Università di Modena e Reggio Emilia, Modena, Italy
[3] Institute for Advanced Study (IAS), University of Amsterdam,
Amsterdam, Netherlands
[4] European Centre for Living Technology, Venice, Italy

Abstract. Boolean networks are currently acknowledged as a powerful model for cell dynamics phenomena. Recently, the possibility of modelling methylation mechanisms—involved in cell differentiation—in Random Boolean Networks have been discussed: methylated genes are represented in the network as nodes locked to value 0 (*frozen nodes*). Preliminary results show that this mechanism can reproduce dynamics with characteristics in agreement with those of cell undergoing differentiation. In a second, parallel work, the effect of nodes with self-loops in Random Boolean Networks has been studied, showing that the average number of attractors may increase with the number of self-loops, whilst the average attractor robustness tends to decrease. As these two studies are aimed at extending the applicability of Random Boolean Networks to model cell differentiation phenomena, in this work we study the combined effect of the previous two approaches. Results in simulation show that frozen nodes tend to partially dampen the effects of self-loops on attractor number and robustness. This outcome suggests that both the variants can indeed be effectively combined in Boolean models for cell differentiation.

Keywords: Random Boolean Networks · Self-loops · Attractors · Chromatin dynamics

1 Introduction

Chromatin[1] dynamics significantly contributes—by modulating access to genes—to differential gene expression, and ultimately to determine cell types. More specifically, methylation processes change the degree of compactness of the chromatin,

---

[1] Condensed structure in which the DNA of eukaryotic cells is organised.

© Springer Nature Switzerland AG 2020
F. Cicirelli et al. (Eds.): WIVACE 2019, CCIS 1200, pp. 28–37, 2020.
https://doi.org/10.1007/978-3-030-45016-8_4

most often making DNA regions inaccessible [7,16,17]. Along differentiation lineages, the attained configurations of DNA methylation are inherited and progressively extended as cells become more specialised [13]. Recently, supported by these evidences of the key role of chromatin modifications in cell differentiation, a model of the effects of the methylation mechanism [6] have been introduced in the dynamics of Boolean models [9] of Genetic Regulatory Networks (GRNs). This mechanism—tested on Random Boolean Network (RBN) ensembles—statistically produces a decrease of the attractor number and a dynamics characterised by behaviours resembling ordered RBNs. However, this mechanism does not reduce variability as it still permits the possibility of generating diverse differentiation paths, i.e. cell types determined by the specific sequence of methylated genes. Recently, a variant of the classical RBN model has been investigated with the aim of including features observed in GRN models of genetic control in real organisms; this variant consists in introducing self-loops in RBNs, i.e. direct gene self-regulation [5,14,15]. These studies show that the gross effect of the introduction of self-loops in RBNs is to increase the number of attractors and to decrease the average probability of returning to an attractor after a transient perturbation.

In this work we study the combined effect of the two above mentioned variants in RBNs. The motivation supporting this study is to extend classical RBN models so as to capture differentiation phenomena more accurately and provide a model suitable for comparisons with real data. On the one hand, the methylation mechanism based on clamping nodes at 0 has the main effect of stabilising the network; on the other hand, self-loops may increase the number of attractors in a RBN[2] and may reduce the average probability of returning to the same attractor after a perturbation. Therefore, the main question is whether these two mechanisms tend to compensate each other and a balanced combination of the individual effects is attained and, if so, under which conditions this happens. In Sect. 2 we introduce the model and we detail the experimental setting. Results are presented and discussed in Sect. 3, and we conclude with Sect. 4.

2   Materials and Methods

Boolean Networks (BNs), introduced by Kauffman [9], have been successfully used as GRN models both for identifying generic properties of statistical ensemble of networks [12] and for reproducing the dynamics of specific reconstructed biological GRNs [4,8]. Random Boolean Networks (RBNs) are the most studied generic ensemble of boolean GRN models: their topology and functions are randomly generated by specifying the number of inputs per node k—excluding the possibility of direct auto-regulation—and the probability p by which value 1 is assigned to an entry of a node truth table [1–3,10]. The study of properties of BNs with self-loops represents a step towards the definition, and refinement, of

---

[2] This happens in particular when the self-regulation is modelled by a canalizing function, such as the logical OR.

Table 1. Summary of the experimental parameters.

Number of nodes (n)	Number of self-loops (N_{sl})	Number of frozen nodes (N_f)
20	0, 3, 6	0, 3, 6, 9
50	0, 5, 10	0, 5, 10, 20

ensembles of more realistic BNs models; these can more accurately capture certain mechanisms underlying cell behaviour [14,15]. The main effect of self-loops in RBNs is to increase, on average, the number of attractors and decrease their robustness. This effect is striking when nodes have 2 inputs (and one is from the node to itself) and a canalizing function, such as the logical OR. In this case, once the dynamics has led the node to assume value 1, this node will assume the same value indefinitely.[3] Therefore, a moderate number of self-regulated nodes may introduce more variability (i.e. more equilibrium states in the system), but a large number might make the network highly unstable.

With the aim of enriching classical RBNs with mechanisms more specifically motivated by modelling cell differentiation, a simplified implementation of methylation mechanisms in RBNs has been introduced, consisting in clamping to value 0 one or more nodes; in the following, we denote those nodes as *frozen* [6]. Note that this mechanism introduces a symmetry breaking in the RBN dynamics, as it attributes a precise meaning to the value 0: frozen nodes are those that cannot be expressed. Preliminary results show that, as the number frozen nodes increases, the network becomes more ordered. Despite this, the possibility of reaching different attractors is still permitted by exerting different freezing sequences. This result suggests that this mechanism may provide an effective contribution to RBN models for cell differentiation.

In this work we combine the freezing mechanism with the self-loop variant and observe their compound effect in RBN ensembles. For all the experiments, statistics are taken across 100 RBNs with $n \in \{20, 50\}$, $k = 2$ and $p = 0.5$. The BNs used in these experiments are subject to a synchronous and deterministic dynamics. In BN models, cell types may be represented by attractors—or sets of attractors, depending on the interpretation chosen. We measure the *robustness* of an attractor as the probability of returning to it after a temporary flip of the value of a randomly chosen node. In our experiments we sampled the possible transitions between attractors after a perturbation and recorded the *returning probability* to each of the attractors sampled. For each network we took the average returning probability computed across all the attractors—omitting the cases with only one attractor. Experiments were performed for a varying number of frozen nodes and number of self-loops N_{sl}, depending on the size of the network (see Table 1). The number of self-loops considered for $n = 50$ had to be limited due to the exponential increase of attractors number, which might make the computation of the attractor returning probability computationally impractical.

---

[3] The same, of course, happens with the AND function and value 0.

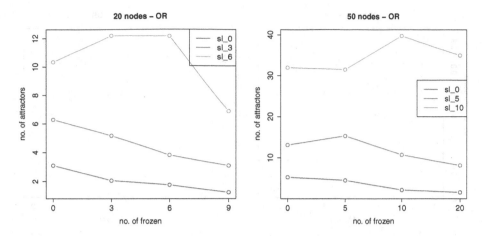

Fig. 1. Mean attractor number in BNs with 20 (left) and 50 (right) nodes for several combinations of number of self-loops and frozen nodes. Self-loops added in OR.

According to previous experimental settings [14], nodes with self-loops are chosen at random and ruled by *OR*, *AND* or random (RND) logical functions (i.e. $p = 0.5$).

3   Results

The first subject of our analysis is the number of attractors. Indeed, attractors play a fundamental role in BN models of differentiation, as attractors—or sets of thereof—are associated to cell types. Moreover, the number of attractors of a BNs is a reckon of the possible equilibrium states the system has and so the possible 'answers' it provides to perturbations: as stated by Kauffman [11], a real cell should be able to discriminate among a number of classes that provides the optimal balance between classification of the environmental stimuli and reliable and robust classification. Figure 1 shows the trend of the average number of attractors in the case of self-loops in OR. We observe that for a fixed number of frozen nodes, the number of attractors increases with self-loops. This result confirms previous results on RBNs with self-loops. The trend at fixed number of self-loops is instead quite informative: frozen nodes tends to compensate the increase of the attractors number. This result is striking in the case of 20 nodes RBNs, whilst for $n = 50$ the effect is less marked even though the containment of the number of attractors is anyway clear.

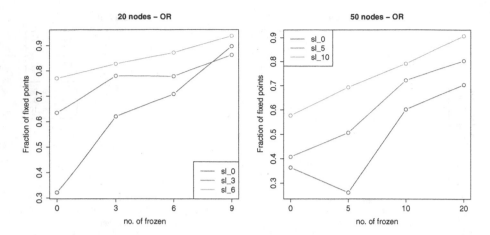

Fig. 2. Mean fraction of fixed points in BNs with 20 (left) and 50 (right) nodes for several combinations of number of self-loops and frozen nodes. Self-loops added in OR.

In previous work, it has also been observed that the fraction of fixed points among the attractors increases with the number of frozen nodes or the number of self-loops. Figure 2 shows the fraction of fixed points, averaged across all the attractors. As expected, the net effect is that fixed points increase both with the number of frozen nodes and the number of self-loops.

These results on the number of attractors are obtained in the case of self-loops in OR, which have the property of keeping indefinitely the value of a node at 1 after it has reached this value during the regular updates of the network. On the other side, freezing a node means setting it to 0 forever. One may ask what is the effect of using different Boolean functions in the nodes with self-loop, in particular in the case of AND functions which, in a sense, impose the same bias as freezing towards zero. Results of this latter case are summarised in Figs. 3 and 4, where we can observe that the combined effect is to drastically reduce the number of attractors. A similar trend, although somehow diluted, is obtained when self-loops are associated to random functions (see Figs. 5 and 6).

Results on average attractor robustness are shown in Fig. 7, in the case of self-loops in OR, where the mean robustness is plotted against N_f for all the values of N_{sl}. If we focus on the trend of a single curve, we observe that the average robustness monotonically increases with the number of frozen nodes. This result is probably not surprising but it is the first time it is experimentally assessed. When we consider the trend as a function of the number of nodes with self-loops we observe a non-monotonic behaviour: a moderate number of self-loops is further reinforcing the effect of frozen nodes, but a greater one has detrimental effects on robustness. This result is in agreement with previous ones [15], where it was shown that a fraction of nodes with self-loops higher than about 20% makes average robustness drop. These results support the hypothesis that

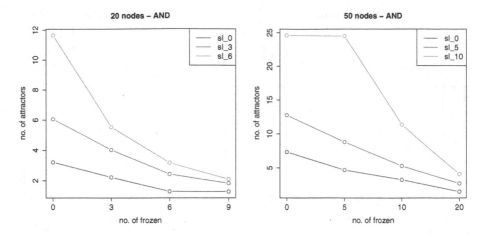

Fig. 3. Mean attractor number in BNs with 20 (left) and 50 (right) nodes for several combinations of number of self-loops and frozen nodes. Self-loops added in AND.

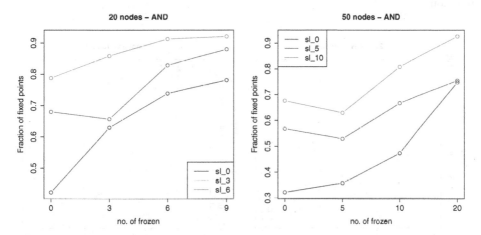

Fig. 4. Mean fraction of fixed points in BNs with 20 (left) and 50 (right) nodes for several combinations of number of self-loops and frozen nodes. Self-loops added in AND.

a mild fraction of nodes with self-loops is beneficial for RBNs attractor robustness, especially if combined with freezing mechanisms that model methylation. Current biological cells are the result of evolution, therefore Boolean models of them are not random; nevertheless, if random BNs with our variants are proven to exhibit features closer to the ones of real cells than simple RBNs, then this enriched RBN model is likely to be a more accurate model of real cells and capture relevant phenomena with higher accuracy. Moreover, in an evolutionary perspective, RBNs with self-loops and methylation mechanisms provide a more suitable starting condition for the evolution of models towards given target

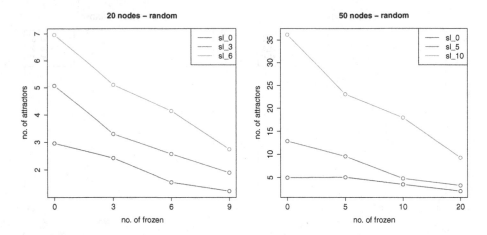

Fig. 5. Mean attractor number in BNs with 20 (left) and 50 (right) nodes for several combinations of number of self-loops and frozen nodes. Self-loops added with a random function.

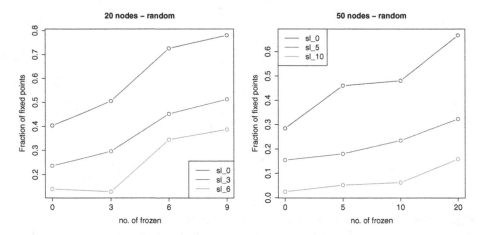

Fig. 6. Mean fraction of fixed points in BNs with 20 (left) and 50 (right) nodes for several combinations of number of self-loops and frozen nodes. Self-loops added with a random function.

characteristics. The case of self-loops with AND function provides a somewhat different picture, because the tension between frozen nodes and self-loops is no longer present and the average robustness tends to increase (see Fig. 8). Analogous considerations hold for self-loops with random functions, even though less marked (see Fig. 9).

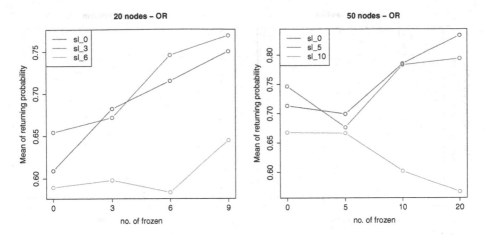

Fig. 7. Mean attractor robustness in BNs with 20 (left) and 50 (right) nodes for several combinations of number of self-loops and frozen nodes. Self-loops added in OR.

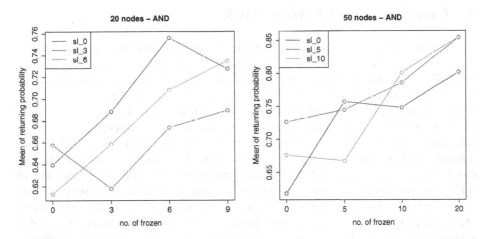

Fig. 8. Mean attractor robustness in BNs with 20 (left) and 50 (right) nodes for several combinations of number of self-loops and frozen nodes. Self-loops added in AND.

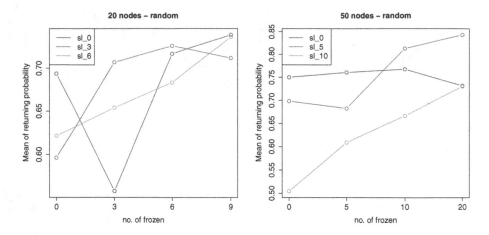

Fig. 9. Mean attractor robustness in BNs with 20 (left) and 50 (right) nodes for several combinations of number of self-loops and frozen nodes. Self-loops added with a random function.

4   Conclusion and Future Work

Boolean networks, and in particular random Boolean networks, have been proven to capture relevant phenomena in cell dynamics. Recently, a variant of the classical RBN model and an enriched dynamical scheme have been introduced: the first consists in considering also self-loops (i.e. direct self-regulation) and the second is a simplified model of methylation consisting in freezing (i.e. clamp to 0) some nodes of the network. In this work we have presented a study on the effects resulting from the combination of these two mechanisms. The main outcome is that frozen nodes dampen the effect of self-loops (mainly in the OR case) that increase the number of attractors; moreover, a moderate amount of self-loops combined with frozen nodes seem to favour attractor robustness. This result suggests the combination of these two variants for modelling differentiation phenomena, in which the possibility of expressing a number of possible attractors, some of which highly stable, is one of the main desired properties.

These results support further investigations, in two main directions. On one side, we plan to identify suitable data from real cell experiments that can be compared with our model. On the other side, as the properties we have observed are related to random models, *in silico* evolutionary experiments are needed to assess whether the property we have observed provide an advantage for evolving GRNs capable of showing the main properties of differentiation.

References

1. Aldana, M., Coppersmith, S., Kadanoff, L.P.: Boolean dynamics with random couplings. In: Kaplan, E., Marsden, J.E., Sreenivasan, K.R. (eds.) Perspectives and Problems in Nolinear Science, pp. 23–89. Springer, New York (2003). https://doi.org/10.1007/978-0-387-21789-5_2
2. Bastolla, U., Parisi, G.: The modular structure of kauffman networks. Physica D: Nonlinear Phenom. **115**(3–4), 219–233 (1998)
3. Bastolla, U., Parisi, G.: Relevant elements, magnetization and dynamical properties in kauffman networks: a numerical study. Physica D **115**(3–4), 203–218 (1998)
4. Bornholdt, S.: Boolean network models of cellular regulation: prospects and limitations. J. R. Soc. Interface **5**(Suppl. 1), S85–S94 (2008)
5. Braccini, M., Montagna, S., Roli, A.: Self-loops favour diversification and asymmetric transitions between attractors in boolean network models. In: Cagnoni, S., Mordonini, M., Pecori, R., Roli, A., Villani, M. (eds.) WIVACE 2018. CCIS, vol. 900, pp. 30–41. Springer, Cham (2019). https://doi.org/10.1007/978-3-030-21733-4_3
6. Braccini, M., Roli, A., Villani, M., Montagna, S., Serra, R.: A simplified model of chromatin dynamics drives differentiation process in Boolean models of GRN. In: The 2019 Conference on Artificial Life (ALIFE) (2019)
7. Gilbert, S.F., Barresi, M.J.F.: Developmental biology, chap. 3, XI edn., pp. 50–52. Sinauer Associates Inc. (2016)
8. Helikar, T., Kowal, B., McClenathan, S., Bruckner, M., Rowley, T., Madrahimov, A., Wicks, B., Shrestha, M., Limbu, K., Rogers, J.A.: The cell collective: toward an open and collaborative approach to systems biology. BMC Syst. Biol. **6**(1), 96 (2012)
9. Kauffman, S.A.: Metabolic stability and epigenesis in randomly constructed genetic nets. J. Theoret. Biol. **22**(3), 437–467 (1969)
10. Kauffman, S.A.: The Origins of Order. Oxford University Press, Oxford (1993)
11. Kauffman, S.A.: Investigations. Oxford University Press, Oxford (2000)
12. Kauffman, S.A.: A proposal for using the ensemble approach to understand genetic regulatory networks. J. Theoret. Biol. **230**(4), 581–590 (2004)
13. Kim, M., Costello, J.: Dna methylation: an epigenetic mark of cellular memory. Exp. Mol. Med. **49**(4), e322 (2017)
14. Montagna, S., Braccini, M., Roli, A.: The impact of self-loops on Boolean networks attractor landscape and implications for cell differentiation modelling. In: IEEE/ACM Trans. Computat. Biol. Bioinform. https://doi.org/10.1109/TCBB.2020.2968310. (Early access)
15. Montagna, S., Braccini, M., Roli, A.: The impact of self-loops in random boolean network dynamics: a simulation analysis. In: Pelillo, M., Poli, I., Roli, A., Serra, R., Slanzi, D., Villani, M. (eds.) WIVACE 2017. CCIS, vol. 830, pp. 104–115. Springer, Cham (2018). https://doi.org/10.1007/978-3-319-78658-2_8
16. Perino, M., Veenstra, G.: Chromatin control of developmental dynamics and plasticity. Dev. Cell **38**(6), 610–620 (2016)
17. Schuettengruber, B., Cavalli, G.: Recruitment of polycomb group complexes and their role in the dynamic regulation of cell fate choice. Development **136**(21), 3531–3542 (2009)

A Memetic Approach
for the Orienteering Problem

Valentino Santucci[1]([⊠])[iD] and Marco Baioletti[2][iD]

[1] Department of Humanities and Social Sciences,
University for Foreigners of Perugia, Perugia, Italy
valentino.santucci@unistrapg.it
[2] Department of Mathematics and Computer Science, University of Perugia,
Perugia, Italy
marco.baioletti@unipg.it

Abstract. In this paper we present a new memetic approach to solve the orienteering problem. The key method of our proposal is the procedure **ReduceExtend** which, starting from a permutation of all the vertices in the orienteering problem, produces a feasible path with a locally optimal score. This procedure is coupled with an evolutionary algorithm which navigate the search space of permutations. In our experiments we have considered the following algorithms: the algebraic differential evolution algorithm, and the three continuous algorithms CMA-ES, DE and PSO equipped with the random key technique. The experimental results show that the proposed approach is competitive with the state of the art results of some selected benchmark instances.

1 Introduction

The Orienteering Problem is an important routing problem widely studied in literature [18]. The aim is to generate a path through a set of given nodes, which would maximize total score and would not exceed the given budget.

Formally, an instance of the Orienteering Problem (OP) is given by means of a complete graph $G = (V, E)$ where $V = \{v_0, \dots, v_{n+1}\}$ is the vertex set and E is the edge set. Moreover, every vertex $v_i \in V$ has a score $s_i \geq 0$. The first and last vertex have null score, i.e., $s_0 = s_{n+1} = 0$. Every edge $e_{ij} \in E$ is marked with a travel time $t_{ij} \geq 0$. Finally, a time budget $T_{\max} > 0$ is also given.

The aim of the OP is to determine a path through some vertices of G whose total travel time is limited by $T_{\max}$ and its total score is maximized. Formally, given the sequence of vertices $p = \langle p_0, p_1, \dots, p_l, p_{l+1} \rangle$, its total travel time $t(p)$ and total score $s(p)$ are defined as follows:

$$t(p) = \sum_{i=0}^{l} t_{p_i, p_{i+1}}; \qquad (1)$$

$$s(p) = \sum_{i=0}^{l+1} s_{p_i}. \qquad (2)$$

© Springer Nature Switzerland AG 2020
F. Cicirelli et al. (Eds.): WIVACE 2019, CCIS 1200, pp. 38–48, 2020.
https://doi.org/10.1007/978-3-030-45016-8_5

Hence, the OP aims at a sequence of vertices $p = \langle p_0, p_1, \ldots, p_l, p_{l+1} \rangle$ such that:

- the first and the last vertexes are fixed to the given start and end vertices, i.e., $p_0 = v_0$ and $p_{l+1} = v_{n+1}$;
- each vertex of V appears at most once in p, i.e., p has to be an Hamiltonian path over a subset of V;
- the total travel time $t(p)$ satisfies inequality (3):

$$t(p) \leq T_{\max}; \tag{3}$$

- the total score $s(p)$ is maximized.

The OP is an NP-hard optimization problem [18, 41]. Moreover, the popular traveling salesman problem (TSP) is a special case of the OP. Indeed, a TSP instance is an OP instance with an infinite budget of time and constant scores on the vertices. For this reason, the search space of the OP is larger than that of the TSP thus, in this sense, solving the OP is computationally more difficult than solving the TSP.

In this paper we propose a novel evolutionary-memetic approach for solving the OP. Evolutionary algorithms are widely popular meta-heuristics which allow to handle computationally difficult optimization problems. They iteratively evolve a population of solutions by means of genetic or swarm-intelligence operators. Often, they are combined with local search methods in order to perform local refinements and to improve their effectiveness. The hybridizations of an evolutionary algorithm with a local search procedure are generally known in literature as memetics algorithms [31]. Here, we introduce a novel heuristic local search procedure for the OP and we combine it with evolutionary algorithms designed to navigate the search space of permutations.

The rest of the paper is organized as follows:

- Section 2 presents a short overview of the OP literature;
- Section 3 describes the main scheme of the approach proposed;
- Section 4 introduces the memetic procedure for the OP;
- Section 5 presents and discusses the experimental results;
- Section 6 concludes this work by summarising the key research outputs and drawing some considerations for possible future developments.

2   Related Work

In this paper we are interested to the classical OP as defined in Sect. 1. Nevertheless, note that other OP variants have been introduced in the literature such as, for example, the team orienteering problem [12] where a fleet of traveling agents is considered, the OP with time windows [24], the time dependent OP [42] where the traveling times are functions of some network's properties, the stochastic OP [43], or the generalized OP [15] where the objective function is a non-linear function.

The classical OP arises in several applications: a traveling salesperson with not enough time to visit all possible cities [40], trucks delivering goods to customers where each customer has a priority and the truck has fuel constraint [17], the single-ring design problem when building telecommunication networks [39], building a mobile tourist guide [37], etc.

In the beginning, several exact algorithms have been proposed to solve the OP. For instance, [25,32] use a branch-and-bound algorithm, [26] introduces a cutting plane method for the OP, and [14,16] further extended the branch-and-cut approach. A classification of the exact algorithms for the OP is presented in [13].

Unfortunately, as for other combinatorial problems, exact approaches allow to handle only relatively small instances. Therefore, meta-heuristics are nowadays the most effective approaches for the OP. Among these: variants of the particle swarm optimization algorithm have been employed in [36,44], a variable neighborhood search approach is adopted in [27], greedy randomized search procedures for the OP are introduced in [11,28], while an evolutionary algorithm for the OP is depicted in [23].

Finally, two survey articles about the OP are [18,41].

3   The Proposed Memetic Approach

We start by noting that, since the first and last vertices of the path to find are fixed to, respectively, v_0 and v_{n+1}, an OP solution can be represented as a permutation of some vertices of $\tilde{V} = V \setminus \{v_0, v_{n+1}\}$.

Let P be the search space of all possible paths from v_0 to v_{n+1}, then P contains all the possible permutations for every subset $U \subseteq \tilde{V}$. Hence,

$$|P| = \sum_{k=0}^{n} \binom{n}{k} k! = (n+1)!$$

Therefore the search space of the OP is larger than that of the more classical permutation-based problems like, for instance, PFSP (Permutation Flowshop Scheduling Problem), QAP (Quadratic Assignment Problem), TSP (Traveling Salesman Problem), or LOP (Linear Ordering Problem).

In the following: we will use the term *full permutation* to refer to a permutation of the whole set $\tilde{V}$, and we denote by S the set of all the full permutations. Note however that S is only a subset of all the possible OP solutions in P, i.e., $S \subset P$.

Our proposal is to tackle the OP by means of two algorithmic components:

1. An evolutionary algorithm $\mathcal{A}$ that navigates the space of full permutations S, and
2. The procedure ReduceExtend which takes in input a full permutation $\pi \in S$ and produces a path $p \in P$ that is feasible under the constraint (3) and whose score $s(p)$ is locally optimal (under some locality definitions as explained in Sect. 4).

$\mathcal{A}$ and ReduceExtend interact with each other by following the memetic principles [31]. $\mathcal{A}$ is the high level scheme which moves in the (smaller) space of full permutations and whose search is guided by a fitness function which embed the ReduceExtend procedure. Indeed, given an individual $\pi \in S$ – in the $\mathcal{A}$'s population – its fitness is computed as $s(\text{ReduceExtend}(\pi))$, i.e., the OP score of the feasible and locally optimal path returned by the ReduceExtend procedure.

This scheme allows to exploit the available evolutionary algorithms (EAs) devised for the permutations search space [1,10,33]. Two main classes of permutation-based EAs are considered:

1. The EAs whose individuals are explicitly represented as full permutations in S, and
2. The EAs which evolve populations formed by real vector individuals and that rely on the random-key procedure in order to decode a real vector into a full permutation [10].

For the first class we have considered the recently proposed Algebraic Differential Evolution for Permutations (ADEP) [6,33], while, for the second class we have chosen the Covariance Matrix Adaptation Evolution Strategy (CMA-ES) [19], the Differential Evolution (DE) [29,38], and the Particle Swarm Optimization (PSO) [22].

The second class algorithms adopts the random key procedure [10] to convert a real vector $x \in \mathbb{R}^n$ to an n-length permutation $\pi \in S$ by replacing each entry x_i of x with the vertex $v_r \in \tilde{V}$ such that r is the ranking of x_i among the other entries of x.

Finally note that, embedding the memetic procedure inside the fitness function of the evolutionary algorithm at hand is referred in the literature as a Baldwinian approach [20], i.e., the individual of the EA is refined through local search but it is kept unmodified in the population of the EA. However, note that, since ADEP directly navigates the space of permutations, it has been possible to also devise a simple procedure which converts a generic OP solution in P to a full permutation in S. Therefore, differently from the second-class approaches, ADEP has also been implemented following the Lamarckian approach [20], i.e., the refined individual is modified in the population of the EA.

4   The ReduceExtend Procedure

The procedure ReduceExtend takes in input a full permutation $\pi \in S$ and produces a path $p \in P$ which is feasible and locally optimal.

Three main subprocedure have been devised: Reduce, Extend, Scramble Extend. They are connected together in the definition of ReduceExtend as follows:

$$\text{ReduceExtend}(\pi) = \text{ScrambleExtend}(\text{Extend}(\text{Reduce}(\pi))), \qquad (4)$$

where π is a given full permutation, i.e., $\pi \in S$ and $\text{ReduceExtend}(\pi)$ is a feasible and locally optimal path in P.

Therefore, `ReduceExtend` acts as a projection from the smaller space S to the larger space P. In the following subsections we describe the three subprocedures `Reduce`, `Extend`, `ScrambleExtend`.

Moreover, Subsect. 4.4 introduces the `PathToFullPermutation` operator that, given a path $p \in P$, stochastically generates a full permutation $\pi = $ `PathToFullPermutation`(p) which is consistent with p. As explained in Sect. 3, this operator allows to devise a Lamarckian variant of the ADEP algorithm.

4.1   Reduce

In this phase, the full permutation $\pi \in S$ is projected into the larger space P by producing a feasible path $p \in P$.

`Reduce` starts with $p = \pi$ and iteratively removes vertices from p till the constraint $t(p) \leq T_{\max}$ is satisfied.

At each iteration, the vertex to be removed is selected to be the one which minimizes the normalized score lost $nsl(p_i)$ defined as the ratio between the score lost and the gain of travel time if the vertex p_i is removed from the path p. Formally,

$$nsl(p_i) = \frac{s(p_i)}{t_{p_{i-1}p_i} + t_{p_i p_{i+1}} - t_{p_{i-1}p_{i+1}}}. \tag{5}$$

Therefore, the `Reduce` procedure modifies the path in input by improving its total travel time $t(p)$ but penalizing its total score $s(p)$. The greedy choice based on the normalized score lost represents a tradeoff between the score lost and the time gain achieved during the iterations of `Reduce`.

4.2   Extend

The greedy behaviour of the `Reduce` procedure may produce a path p which can be extended-back without violating the $T_{\max}$ constraint. Therefore, the procedure `Extend` iteratively inserts a vertex into p until the constraint is violated.

At each iteration, the vertex v – to insert in p – is selected to be the one with the largest score among the vertices not already belonging to p and such that, when inserted into p, does not violate the constraint. The selected vertex is inserted at the position of p which minimizes the travel time increase.

Therefore, oppositely with respect to `Reduce`, the `Extend` procedure modifies the path in input by improving its total score $s(p)$ but penalizing its total travel time $t(p)$. Moreover, the vertex and position selection criteria guarantee a good tradeoff between the score gain and the time lost achieved during the iterations of `Extend`.

4.3   ScrambleExtend

`ScrambleExtend` aims to improve the score of the path p by iteratively alternating two local search steps as done in variable neighborhood search procedures [30].

The two refinement steps consist of a 2OPT search [21], and one step of the Extend procedure previously described.

The 2OPT search scrambles the vertices in p aiming to reduce the travel time $t(p)$ without decreasing the score $s(p)$. Indeed, no vertex is introduced or removed from p, they are only swapped.

The 2OPT local search is widely adopted in the TSP [21]: given a path p it iteratively selects the best neighboring path till a local optimum is reached. Neighbors are defined by means of 2OPT moves, i.e., the path p' is a neighbor of the path p if it is obtained by replacing two arcs in p with two arcs not in p (but using the same vertices of p). As described in [8], 2OPT moves are equivalent to reversing chunks of a given path. Therefore, the 2OPT neighborhood of p has size $\binom{|p|}{2}$.

Note that the 2OPT local search has the effect of decreasing the travel time $t(p)$, thus creating room for the insertion of a new vertex. For this reason, every 2OPT local search is followed by a single step of the Extend procedure (described in Sect. 4.2) which can possibly increase the score.

The iterations of ScrambleExtend terminate when it is no more possible to improve the score of the incumbent path. Note that, ScrambleExtend does not remove any vertex from the input path p, thus it can only increase its score $s(p)$. Anyway, the original ordering of vertices may be modified by the 2OPT steps.

4.4 PathToFullPermutation

Given a feasible path $p \in P$, $PathToFullPermuation(p)$ generates a full permutation $\pi \in S$ which can be used to modify the individual of the evolutionary algorithm at hand following a Lamarckian style of evolution [20].

In order to promote the population diversity in the Lamarckian algorithm, PathToFullPermutation has been designed by following a simple stochastic mechanism: it iteratively insert a randomly selected vertex not yet in p in a random position of p till p is a full permutation.

Despite the stochastic approach, the items in the original p maintain the same relative order among them in the final full permutation.

5   Experiments

Experiments have been held using the 89 instances from the five benchmark suites listed in Table 1 and obtained from the website https://www.mech. kuleuven.be/en/cib/op.

Five EAs have been considered: ADEP-B (Baldwinian), ADEP-L (Lamarckian), CMA-ES, DE and PSO. Note that, as described in Sect. 3 the latter three algorithms are equipped with the random-key decoding scheme and follows the Baldwinian approach.

All the five algorithms have been run 10 times per instance and every execution terminates as soon as the (known) optimal score has been reached or when 10, 000 evaluations have been performed. Therefore, the goal of this experimentation is to investigate how many times and how quickly the optimum is reached.

Table 1. Benchmarks

Benchmark	#Instances	Size
Tsiligirides_set1	18	31
Tsiligirides_set2	11	21
Tsiligirides_set3	20	32
Chao_set64	14	64
Chao_set66	26	66

Table 2 shows two data for each algorithm and every benchmark suites: the number of instances that have been solved in all the 10 executions, and the average number of fitness evaluations performed by the algorithm. Obviously, the first measure has to be maximized while the second one has to minimized. The last row of the table averages the results across the different benchmarks.

Table 2. Experimental results

Benchmark	ADEP-B	ADEP-L	CMA-ES	DE	PSO
Tsiligirides_set1	18/18	18/18	18/18	18/18	18/18
	87	**34**	95	79	114
Tsiligirides_set2	11/11	11/11	11/11	11/11	11/11
	35	**24**	42	28	32
Tsiligirides_set3	20/20	20/20	20/20	20/20	20/20
	30	**19**	28	29	26
Chao_set64	14/14	14/14	14/14	14/14	13/14
	466	**149**	728	618	685
Chao_set66	24/26	**26/26**	18/26	17/26	18/26
	1439	**260**	2220	2605	2297
Average	87/89	**89/89**	81/89	80/89	80/89
	411	**97**	623	672	631

Interestingly, all the algorithms have been able to reach the optimum in every instance at least once. Moreover, most of the times the optimum has been obtained in every execution and by employing a small number of fitness evaluations. These results clearly validates our memetic approach.

Regarding the differences among the different schemes, the Lamarckian variant of ADEP clearly outperforms all the other schemes. Moreover, also the Baldwinian ADEP has been able to outperform the random-key-based CMA-ES, DE and PSO. This clearly suggests that, though the literature presents a large variety of numerical schemes equipped with the random key decoding scheme, the algorithms purposely defined to navigate a combinatorial search space, as ADEP, have to be preferred (at least in the OP case).

6 Conclusion and Future Work

In this paper we have introduced a novel memetic approach for the Orienteering Problem.

In particular, the main contribution is the `ReduceExtend` procedure which transforms any full permutation into a locally optimal feasible solution for the OP. `ReduceExtend` is composed by three subprocedures purposely designed to maximize the objective function by also satisfying the time constraint.

This memetic procedure allows to handle the OP by using any evolutionary algorithm which works on the permutation search space. In this work, we have adopted and experimentally compared five algorithms. In one case it has been possible to use the Lamarckian evolution by introducing a randomized procedure for converting back a feasible OP solution into a full permutation of items. This algorithm, namely ADEP-L, resulted to be the most effective in the experiments we carried out.

Moreover, all the five approaches were able to reach the known optimal values, thus validating our proposal.

As future work, we expect to provide an algebraic formalization of our procedure using the algebraic framework described in [2,4,7,9,34,35]. This framework can also be used to handle the OP by using the concept of product group as done in [5]. Finally, the memetic approach can be extended to other algorithms like, for example, the ant colony optmization for the permutation space [3].

Acknowledgement. The research described in this work has been partially supported by: the research grant "Fondi per i progetti di ricerca scientifica di Ateneo 2019" of the University for Foreigners of Perugia under the project "Algoritmi evolutivi per problemi di ottimizzazione e modelli di apprendimento automatico con applicazioni al Natural Language Processing"; and by RCB-2015 Project "Algoritmi Randomizzati per l'Ottimizzazione e la Navigazione di Reti Semantiche" and RCB-2015 Project "Algoritmi evolutivi per problemi di ottimizzazione combinatorica" of Department of Mathematics and Computer Science of University of Perugia.

References

1. Baioletti, M., Milani, A., Santucci, V.: Algebraic particle swarm optimization for the permutations search space. In: Proceedings of 2017 IEEE Congress on Evolutionary Computation (CEC 2017), pp. 1587–1594 (2017)
2. Baioletti, M., Milani, A., Santucci, V.: Algebraic crossover operators for permutations. In: 2018 IEEE Congress on Evolutionary Computation (CEC 2018), pp. 1–8 (2018). https://doi.org/10.1109/CEC.2018.8477867

3. Baioletti, M., Milani, A., Santucci, V.: A new precedence-based ant colony optimization for permutation problems. In: Shi, Y., Tan, K.C., Zhang, M., Tang, K., Li, X., Zhang, Q., Tan, Y., Middendorf, M., Jin, Y. (eds.) SEAL 2017. LNCS, vol. 10593, pp. 960–971. Springer, Cham (2017). https://doi.org/10.1007/978-3-319-68759-9_79

4. Baioletti, M., Milani, A., Santucci, V.: Automatic algebraic evolutionary algorithms. In: Pelillo, M., Poli, I., Roli, A., Serra, R., Slanzi, D., Villani, M. (eds.) WIVACE 2017. CCIS, vol. 830, pp. 271–283. Springer, Cham (2018). https://doi.org/10.1007/978-3-319-78658-2_20

5. Baioletti, M., Milani, A., Santucci, V.: Learning Bayesian networks with algebraic differential evolution. In: Auger, A., Fonseca, C.M., Lourenço, N., Machado, P., Paquete, L., Whitley, D. (eds.) PPSN 2018. LNCS, vol. 11102, pp. 436–448. Springer, Cham (2018). https://doi.org/10.1007/978-3-319-99259-4_35

6. Baioletti, M., Milani, A., Santucci, V.: MOEA/DEP: an algebraic decomposition-based evolutionary algorithm for the multiobjective permutation flowshop scheduling problem. In: Liefooghe, A., López-Ibáñez, M. (eds.) EvoCOP 2018. LNCS, vol. 10782, pp. 132–145. Springer, Cham (2018). https://doi.org/10.1007/978-3-319-77449-7_9

7. Baioletti, M., Milani, A., Santucci, V.: Variable neighborhood algebraic differential evolution: an application to the linear ordering problem with cumulative costs. Inf. Sci. **507**, 37–52 (2020). https://doi.org/10.1016/j.ins.2019.08.016, http://www.sciencedirect.com/science/article/pii/S0020025519307509

8. Baioletti, M., Milani, A., Santucci, V., Bartoccini, U.: An experimental comparison of algebraic differential evolution using different generating sets. In: Proceedings of the Genetic and Evolutionary Computation Conference Companion, GECCO 2019, pp. 1527–1534. ACM, New York (2019). https://doi.org/10.1145/3319619.3326854, http://doi.acm.org/10.1145/3319619.3326854

9. Baioletti, M., Milani, A., Santucci, V.., Tomassini, M.: Search moves in the local optima networks of permutation spaces: the QAP case. In: Proceedings of the Genetic and Evolutionary Computation Conference Companion, GECCO 2019, pp. 1535–1542. ACM, New York (2019). http://doi.acm.org/10.1145/3319619.3326849

10. Bean, J.C.: Genetic algorithms and random keys for sequencing and optimization. ORSA J. Comput. **6**(2), 154–160 (1994)

11. Campos, V., Martí, R., Sánchez-Oro, J., Duarte, A.: Grasp with path relinking for the orienteering problem. J. Oper. Res. Soc. **65**(12), 1800–1813 (2014). https://doi.org/10.1057/jors.2013.156

12. Chao, I.M., Golden, B.L., Wasil, E.A.: The team orienteering problem. Eur. J. Oper. Res. **88**(3), 464–474 (1996). https://doi.org/10.1016/0377-2217(94)00289-4, http://www.sciencedirect.com/science/article/pii/0377221794002894

13. Feillet, D., Dejax, P., Gendreau, M.: Traveling salesman problems with profits. Transp. Sci. **39**(2), 188–205 (2005)

14. Fischetti, M., González, J.J.S., Toth, P.: Solving the orienteering problem through branch-and-cut. INFORMS J. Comput. **10**(2), 133–148 (1998). https://doi.org/10.1287/ijoc.10.2.133

15. Geem, Z.W., Tseng, C.-L., Park, Y.: Harmony search for generalized orienteering problem: best touring in China. In: Wang, L., Chen, K., Ong, Y.S. (eds.) ICNC 2005. LNCS, vol. 3612, pp. 741–750. Springer, Heidelberg (2005). https://doi.org/10.1007/11539902_91

16. Gendreau, M., Laporte, G., Semet, F.: A branch-and-cut algorithm for the undirected selective traveling salesman problem. Networks **32**(4), 263–273 (1998)

17. Golden, B.L., Levy, L., Vohra, R.: The orienteering problem. Naval Res. Logist. (NRL) **34**(3), 307–318 (1987)
18. Gunawan, A., Lau, H.C., Vansteenwegen, P.: Orienteering problem: a survey of recent variants, solution approaches and applications. Eur. J. Oper. Res. **255**(2), 315–332 (2016)
19. Hansen, N., Muller, S., Koumoutsakos, P.: Reducing the time complexity of the derandomized evolution strategy with covariance matrix adaptation (CMA-ES). Evol. Comput. **11**(1), 1–18 (2003)
20. Hart, W.E., Krasnogor, N., Smith, J.E.: Memetic evolutionary algorithms. In: Hart, W.E., Smith, J.E., Krasnogor, N. (eds.) Recent Advances in Memetic Algorithms. STUDFUZZ, pp. 3–27. Springer, Heidelberg (2005). https://doi.org/10.1007/3-540-32363-5_1
21. Helsgaun, K.: General k-opt submoves for the Lin-Kernighan TSP heuristic. Math. Program. Comput. **1**(2), 119–163 (2009)
22. Kennedy, J., Eberhart, R.: Particle swarm optimization. In: Proceedings of IEEE International Conference on Neural Networks, vol. 4, pp. 1942–1948 (1995)
23. Kobeaga, G., Merino, M., Lozano, J.A.: An efficient evolutionary algorithm for the orienteering problem. Comput. Oper. Res. **90**, 42–59 (2018). https://doi.org/10.1016/j.cor.2017.09.003, http://www.sciencedirect.com/science/article/pii/S0305054817302241
24. Labadie, N., Mansini, R., Melechovský, J., Calvo, R.W.: The team orienteering problem with time windows: an LP-based granular variable neighborhood search. Eur. J. Oper. Res. **220**(1), 15–27 (2012). https://doi.org/10.1016/j.ejor.2012.01.030, http://www.sciencedirect.com/science/article/pii/S0377221712000653
25. Laporte, G., Martello, S.: The selective travelling salesman problem. Discrete Appl. Math. **26**(2), 193–207 (1990). https://doi.org/10.1016/0166-218X(90)90100-Q, http://www.sciencedirect.com/science/article/pii/0166218X9090100Q
26. Leifer, A.C., Rosenwein, M.B.: Strong linear programming relaxations for the orienteering problem. Eur. J. Oper. Res. **73**(3), 517–523 (1994). https://doi.org/10.1016/0377-2217(94)90247-X, http://www.sciencedirect.com/science/article/pii/037722179490247X
27. Liang, Y.C., Kulturel-Konak, S., Lo, M.H.: A multiple-level variable neighborhood search approach to the orienteering problem. J. Ind. Prod. Eng. **30**(4), 238–247 (2013). https://doi.org/10.1080/21681015.2013.818069
28. Marinakis, Y., Politis, M., Marinaki, M., Matsatsinis, N.: A memetic-GRASP algorithm for the solution of the orienteering problem. In: Le Thi, H.A., Pham Dinh, T., Nguyen, N.T. (eds.) Modelling, Computation and Optimization in Information Systems and Management Sciences. AISC, vol. 360, pp. 105–116. Springer, Cham (2015). https://doi.org/10.1007/978-3-319-18167-7_10
29. Milani, A., Santucci, V.: Asynchronous differential evolution. In: IEEE Congress on Evolutionary Computation, pp. 1–7, July 2010. https://doi.org/10.1109/CEC.2010.5586107
30. Mladenović, N., Hansen, P.: Variable neighborhood search. Comput. Oper. Res. **24**(11), 1097–1100 (1997)
31. Moscato, P., Cotta, C.: A gentle introduction to memetic algorithms. In: Glover, F., Kochenberger, G.A. (eds.) Handbook of Metaheuristics. International Series in Operations Research & Management Science, vol. 5, pp. 105–144. Springer, Boston (2003)
32. Ramesh, R., Yoon, Y.S., Karwan, M.H.: An optimal algorithm for the orienteering tour problem. ORSA J. Comput. **4**(2), 155–165 (1992). https://doi.org/10.1287/ijoc.4.2.155

33. Santucci, V., Baioletti, M., Milani, A.: Algebraic differential evolution algorithm for the permutation flowshop scheduling problem with total flowtime criterion. IEEE Trans. Evol. Comput. **20**(5), 682–694 (2016)
34. Santucci, V., Baioletti, M., Milani, A.: Tackling permutation-based optimization problems with an algebraic particle swarm optimization algorithm. Fundam. Inf. **167**(1–2), 133–158 (2019). https://doi.org/10.3233/FI-2019-1812
35. Santucci, V., Baioletti, M., Di Bari, G., Milani, A.: A binary algebraic differential evolution for the multidimensional two-way number partitioning problem. In: Liefooghe, A., Paquete, L. (eds.) EvoCOP 2019. LNCS, vol. 11452, pp. 17–32. Springer, Cham (2019). https://doi.org/10.1007/978-3-030-16711-0_2
36. Sevkli, Z., Sevilgen, F.E.: Discrete particle swarm optimization for the orienteering problem. In: IEEE Congress on Evolutionary Computation, pp. 1–8, July 2010. https://doi.org/10.1109/CEC.2010.5586532
37. Souffriau, W., Vansteenwegen, P., Berghe, G.V., Oudheusden, D.V.: A path relinking approach for the team orienteering problem. Comput. Oper. Res. **37**(11), 1853–1859 (2010). https://doi.org/10.1016/j.cor.2009.05.002, http://www.sciencedirect.com/science/article/pii/S0305054809001464, metaheuristics for Logistics and Vehicle Routing
38. Storn, R., Price, K.: Differential evolution - a simple and efficient heuristic for global optimization over continuous spaces. J. Global Optim. **11**(4), 341–359 (1997). https://doi.org/10.1023/A:1008202821328
39. Thomadsen, T., Stidsen, T.: The quadratic selective travelling salesman problem. Informatics and mathematical modelling technical report 2003–17. Technical University of Denmark (2003)
40. Tsiligirides, T.: Heuristic methods applied to orienteering. J. Oper. Res. Soc. **35**(9), 797–809 (1984). https://doi.org/10.1057/jors.1984.162, https://doi.org/10.1057/jors.1984.162
41. Vansteenwegen, P., Souffriau, W., Oudheusden, D.V.: The orienteering problem: a survey. Eur. J. Oper. Res. **209**(1), 1–10 (2011)
42. Verbeeck, C., Sörensen, K., Aghezzaf, E.H., Vansteenwegen, P.: A fast solution method for the time-dependent orienteering problem. Eur. J. Oper. Res. **236**(2), 419–432 (2014). https://doi.org/10.1016/j.ejor.2013.11.038, http://www.sciencedirect.com/science/article/pii/S0377221713009557
43. Ílhan, T., Iravani, S.M.R., Daskin, M.S.: The orienteering problem with stochastic profits. IIE Trans. **40**(4), 406–421 (2008). https://doi.org/10.1080/07408170701592481
44. Şevkli, A., Sevilgen, F.: Stpso: Strengthened particle swarm optimization. Turk. J. Electr. Eng. Comput. Sci. **18**(6), 1095–1114 (2010). https://doi.org/10.3906/elk-0909-18, https://www.scopus.com/inward/record.uri?eid=2-s2.0-78649355029&doi=10.3906%2felk-0909-18&partnerID=40&md5=74766ce1bd8c8970b3d6343b5f23e4a6

The Detection of Dynamical Organization in Cancer Evolution Models

Laura Sani[1], Gianluca D'Addese[2], Alex Graudenzi[3], and Marco Villani[2,4]($\boxtimes$)

[1] Department of Engineering and Architecture, University of Parma, Parma, Italy
[2] Department of Physics, Informatics and Mathematics,
University of Modena and Reggio Emilia, Modena, Italy
`marco.villani@unimore.it`
[3] Institute of Molecular Bioimaging and Physiology of the Italian National Research
Council (IBFM-CNR), Segrate, Milan, Italy
[4] European Centre for Living Technology, Venice, Italy

Abstract. Many systems in nature, society and technology are composed of numerous interacting parts. Very often these dynamics lead to
the formation of medium-level structures, whose detection could allow a
high-level description of the dynamical organization of the system itself,
and thus to its understanding. In this work we apply this idea to the
"cancer evolution" models, of which each individual patient represents
a particular instance. This approach - in this paper based on the RI
methodology, which is based on entropic measures - allows us to identify
distinct independent cancer progression patterns in simulated patients,
planning a road towards applications to real cases.

Keywords: Complex systems analysis · Information theory · Cancer
evolution · Relevance index

1 Introduction

Many systems in nature, society and technology are composed of numerous interacting parts [1,14]. A big part of these systems is characterized by groups of variables that appear to be well coordinated among themselves and have a weaker
interaction with the remainder of the system (Relevant Sets, or shortly RSs in
the following). The capacity of detecting their presence can often lead to a high-
level description of the dynamical organization of a complex system, and thus to
its understanding [14]. Notable examples are the identification of functional neu-
ronal regions in the brain [31], autocatalytic systems in chemistry [13,34,35], the
identification of communities in socio-economic systems [2,12], and the detection
of specific groups of genes ruling the dynamics of a genetic network [35,36]. The
last case is particularly interesting, presenting it a type of analysis based on the
juxtaposition of dynamic states of the system without necessarily following a
temporal order. In other words, in these cases the researchers infer information
about the organization of a system starting from the observation of some of the

© Springer Nature Switzerland AG 2020
F. Cicirelli et al. (Eds.): WIVACE 2019, CCIS 1200, pp. 49–61, 2020.
https://doi.org/10.1007/978-3-030-45016-8_6

dynamic behaviors the organization is able to support. In this work we apply this idea to the "tumor system", of which each individual patient represents a particular instance. The aim is therefore to infer information on the dynamic organization of the tumor process starting from the observation of the damages undergone by a group of patients suffering from the same type of disease. In particular, we are interested in identifying relationships between genome mutations that lead to tumor progression, in order to recognize the presence of distinct cancer progression patterns.

The present work is based on the Relevance Index (RI) metrics [34,35], a set of information-theoretical metrics for the analysis of complex systems that can be used to detect the main interacting structures (RSs) within them, starting from the observation of the status of the system variables over time. The RI methodology is able to provide an effective description of different kinds of systems and can be applied to a broad range of non-stationary dynamical systems, as shown in [24,25,28,36].

The rest of the paper is structured as follows. Section 2 presents the biological context of the present work. Section 3 provides a brief review of the RI metrics, with a particular focus on the zI metric. Section 4 shows our application of the RI method to cancer progression, seen as a particular biological system. Finally, Sect. 5 draws some conclusions, discussing future research directions.

2   Identification of Independent Progressions in Cancer Evolution Models

Cancer is an evolutionary process according to which cancer cells progressively accumulate distinct (epi)genomic alterations (e.g., single-nucleotide variants, SNVs, and/or copy number alterations, CNAs, etc.) some of which - the so-called *drivers* - provide a selective advantage to the cells, which live and proliferate in a complex microenvironment with limited resources [17,20]. As a result of this complex interplay that involves a large and varying number of competing cancer subpopulations, high levels of inter- and intra-tumor heterogeneity are observed in most tumor (sub)types. Such heterogeneity is the major cause of drug resistance, treatment failure and relapses [5,19,38].

For this reason, in recent years a large number of statistical and machine-learning approaches is being developed to take full advantage of the impressive amount of *omics* data produced by massively parallel sequencing of cancer samples [3]. The goal is to decipher and characterize the somatic evolution of tumors, in order to identify possible regularities and differences between cancer (sub)types and to possibly isolate the molecular "weak points" in the evolutionary trajectories (e.g., bottlenecks), which may be targeted with ad-hoc therapeutic strategies [10].

In [6] some of the authors have introduced a computational pipeline (named PICNIC) for the reconstruction of cancer evolution models from cross-sectional somatic mutation profiles of multiple cancer patients. The theoretical framework combines Suppes' notion of *probabilistic causality* [30] and maximum likelihood estimation, in order to extract useful information on the underlying evolutionary process from noisy and often limited data [15,22,23]. The final outcome of the pipeline is a probabilistic graphical model which highlights the most likely accumulation paths of somatic mutations that characterize a specific cancer cohort. Not only such a model delivers an explanatory model of the evolution of a tumor type and of its heterogeneity, by highlighting the evolutionary paths that characterize distinct patients, but it also allows one to generate predictions, by identifying the most likely future stages of the tumor progression, which may be targeted in useful time. The application to various cancer datasets from The Cancer Genome Atlas (TCGA) allowed to automatically generate experimental hypotheses with translational relevance [18].

More in detail, one of the main steps of the pipeline is the stratification of patients in homogenous subgroups on the basis of their molecular makeup. On the one hand, a correct stratification should allow one to reduce the possible impact of confounding factors of heterogeneous cohorts on the inference accuracy [11]. On the other hand, a statistically robust stratification might be effective in the definition of personalized treatments, as a unique therapy might have even very different efficacy on distinct patients, due to the molecular heterogeneity of the tumor composition.

In certain cases, stratification of patients can benefit from known clinical biomarkers [4], but in most cases it is necessary to employ unsupervised clustering techniques, such as non-negative matrix factorization (NMF) [9] or even classical methods such as k-means or Gaussian mixtures, usually on gene expression data [16]. In general, a successful stratification should identify disjoint sets of patients affected by distinct cancer subtypes, whose evolutionary history, i.e., the genomic alteration accumulation process, is significantly different among one another. Conversely, without stratification, inference methods such as PICNIC might struggle or even produce wrong outcome models, due to the confounding effects that are inherent of heterogeneous and intermixed cohorts (an effect commonly known as the Simpson paradox). The RI methodology allows one to identify groups of variables, therefore it could constitute a useful and interesting pre-processing tool.

3   The RI Methodology

The observed cancer progression can be considered as a "complex system" in which each patient is a particular realization of this dynamical process. Thus, the RI methodology can be applied to identify the "organized groups" of mutations, i.e., the distinct trees describing independent tumour progression, which might be used, in turn, to stratify the patients in homogenous subgroups.

In this section we provide an overview of the theoretical background of the RI metrics, a set of information-theoretical measures that can be used to dynamically analyze complex systems.

3.1 Relevance Index (RI) Metrics

The RI metrics allow one to identify the main interacting structures within a dynamical system. The purpose of the RI metrics is the identification of subsets of the system variables that behave in a coordinated way. This means that the variables belonging to the subset are integrated with each other much more tightly than with the other variables of the system. These subsets can be used to describe the whole system organization, thus they are named Relevant Subsets (RSs).

The RI metrics are information theoretical measures based on Shannon's *entropy* [7]. Their computation relies on the approximation of the statistical distribution of the variables, based on the analysis of a sample of the system states observed over a given time interval. The probabilities are estimated as the relative frequencies of the values observed for each variable. Then, a relevance score is assigned to each group of variables: the higher the score, the more relevant the group. The reference method from which the RI metrics derive is the *Cluster Index* (CI), which was introduced by Tononi et al. [31,32] in the study of biological neural networks close to a stationary state. In [34,35], the CI has been generalized to study dynamical systems, in order to apply the method to a broad range of systems, introducing the *Relevance Index* (RI) metric.

Let us consider a system U described by N discrete variables whose status changes in time, and let us suppose the time series of their values is available. Given a subset S_k composed of k variables, with $k < N$, the RI is defined as the ratio between the *integration* (I, to be maximized), which measures the mutual dependence among the k elements in S_k, and the *mutual information* (MI, to be minimized), which measures the mutual dependence between the subset S_k and the rest of the system $U \setminus S_k$:

$$I(S_k) = \sum_{s \in S_k} H(s) - H(S_k) \tag{1}$$

$$MI(S_k; U \setminus S_k) = H(S_k) + H(U \setminus S_k) - H(S_k, U \setminus S_k) \tag{2}$$

where $H(X)$ is the entropy or the joint entropy, depending on X being a single random variable or a set of random variables.

I, MI and RI, and their normalized counterparts, which solve some issues with their dependency on group size, constitute a family of information-theoretical metrics that can be denoted as RI metrics, since they assess the relevance of a subset of variables for the description of the system organization.

In many cases, the sole integration seems to provide by itself the majority of the relevant information regarding the tightness of subset variables. In the following subsection, we describe the zI metric, which is based on the observation that the integration, multiplied by twice the number of observations, approximately follows a chi-squared distribution with degrees of freedom depending on the number of variables belonging to the analyzed subgroup and on the cardinality of their alphabets [27].

3.2   The zI Metric

The zI metric is defined as follows, where n is the number of observations or instances of the whole system, S_k is a subset of k out of N variables and Sh_k is a subset of dimension k extracted from a homogeneous system U_h (i.e., a system with the same number of variables of the analyzed system and no dynamical organization, with all N variables being independent):

$$zI(S_k) = \frac{2nI(S_k) - \langle 2nI(Sh_k) \rangle}{\sigma(2nI(Sh_k))} \tag{3}$$

As shown in [27], $2nI$ is chi-squared distributed [21]:

$$2nI \approx \chi_d^2 \tag{4}$$

under the null hypothesis of independent random variables and with n large enough.

Moreover, it can be demonstrated that, considering a subset of size k, the degrees of freedom for Eq. 4 can be expressed as:

$$d = (\prod_{j=1}^{k} L_j - 1) - (\sum_{j=1}^{k}(L_j - 1)) \tag{5}$$

where L_j is the cardinality of the alphabet of the subset variable x_j.

Furthermore, it is known [21] that the mean μ and the standard deviation σ of the chi-squared distribution with d degrees of freedom are

$$\mu = d \tag{6}$$
$$\sigma = \sqrt{2d} \tag{7}$$

Thus, the elements of Eq. 3 are all defined and can be simply computed without relying on onerous computational techniques to "simulate" the homogeneous system distribution and to compute the statistics for each subset dimension.

3.3   Iterative Sieving Algorithm

In order to identify the most relevant information, an iterative sieving algorithm has been proposed [37].

The sieving algorithm is based on the consideration that if a set A is a proper subset or superset of a set B and ranks higher than B, then A should be considered more relevant than B. Therefore, the sieve keeps only those sets that are not included in or do not include any other set with higher zI.

In order to analyze the hierarchical organization of a complex system, we proposed an iterative version of the sieving method, that groups one or more sets into a single entity to derive a hierarchy. The simplest, yet effective, way to do so consists in iteratively running the sieving algorithm on the same data, each time using a new representation, in which the top-ranked RS of the previous iteration, in terms of zI values, is considered as atomic and is substituted by a single variable (group variable). Each run produces a new atomic group of variables composed of both single variables and group variables introduced in previous iterations. The iterations of the sieving algorithm come to an end when the value of the considered RI metric (i.e., the zI metric) of the top-ranked RS falls below a pre-set threshold, usually equal to 3.0 (i.e., 3 standard deviations from the reference condition of variable independence). The iterative sieving approach is thus able to highlight the organization of a complex system into sets of variables, which interact with one another at different hierarchical levels, detected, in turn, in the different iterations of the sieve.

The computation of the RI metrics, which is the basic step of the iterative sieving algorithm, is a rather lengthy procedure. Therefore, an efficient implementation becomes a mandatory requirement. The computation time needed to run the method can be reduced, on the one hand, by implementing the zI computation algorithm as massively-parallel GPU code [33], while, on the other hand, by designing some Niching metaheuristics [26,29] to address the search toward the basins of attraction of the main local maxima.

4   Experimental Results

4.1   Generation of Test Cases

In order to assess the efficacy and robustness of our method in identifying the sets of variables corresponding to independent cancer progressions, here we rely on simulated data in a variety of in-silico scenarios.

In this case, we sampled a large number of synthetic binary datasets describing cross-sectional mutational profiles of cancer patients from distinct generative topologies. Each generative topology is described via a probabilistic graphical model, in which each node represents a genomic alteration and each edge corresponds to an evolutionary trajectory (i.e., the process of mutation accumulation) and is characterized by a specific conditional probability (e.g., the edge $A \rightarrow B$ indicates that mutation B can occur with probability $P(B|A)$ and only if A has occurred).

In particular, we here employed *forest* generative topologies - a forest being a disjoint union of trees, with multiple roots - to reflect the assumption of having distinct progression models corresponding to different patient subgroups/cancer subtypes, one per constituting tree. Each tree then describes a different pattern of *branching evolution* [8].

Each binary dataset is represented by a data matrix having n (samples/patients) rows $\times$ m (genomic alterations) columns: each entry is equal to 1 if a certain mutation is present in a sample, 0 otherwise. For each generative topology, binary datasets were sampled starting from the root, with a recursive procedure, and according to randomly assigned conditional probabilities (by ensuring that each mutation is observed at least once in the dataset).

In order to account for noise in the data, due to experimental and technical errors, we finally introduced a parameter $\nu \in (0,1)$ which represents the probability of each entry in the dataset to be random (i.e., false positives $(\epsilon+)$ = false negatives $(\epsilon-) = \nu/2$). The noise level, even if uniformly distributed, complicates the inference problem (see also Fig. 1 for an overall view).

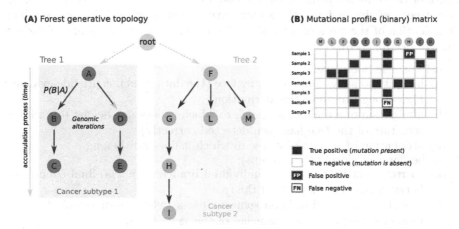

Fig. 1. Synthetic data generation. (A) A number of forest topologies is randomly generated as probabilistic graphical models, in order to model independent tumor progressions (i.e., composed of disjoint sub-models involving different sets of events). Nodes in the graphical models are (epi)genomic alterations and edges represent the accumulation paths. Each edge is characterized by a randomly assigned conditional probability. A fictitious root is included for graphical purposes. The two distinct trees represent independent tumor progressions/subtypes. (B) A large number of independent binary datasets is sampled from the forest generative topologies (1: mutation is present, 0: mutation is absent). False positives and false negatives are included in the datasets depending on the parameters of the simulation.

To summarize our approach, we randomly generated 15 forest topologies with $m = 20$ nodes, and in particular: 5 topologies with 2 roots (i.e., 2 disjoint trees), 5 topologies with 3 roots and 5 topologies with 4 roots. For each topology, we sampled binary datasets with $n = 200$ samples and 3 noise levels: $\nu = 0$, 0.1, and 0.25.

The identification of the distinct trees composing the generative forest topology is extremely relevant, as each tree might be responsible for the evolutionary history of a specific subset of patients and, accordingly, might identify a specific cancer subtype and its evolutionary history.

By associating samples/patients to the trees, e.g., via maximum likelihood approaches, it would be finally possible to produce an effective stratification which might be in turn used in cancer evolution pipelines such as PICNIC, to produce a reliable progression model for each distinct cancer subtype.

4.2 Analysis of Test Cases

We analyzed the forty-five datasets using the RI methodology (using the zI index, and iterating the sieving algorithm until the index fell below the arbitrary reference threshold of 3.0).

The generative forests can be composed by very heterogeneous trees, and the length of the branches inside each tree could be very dissimilar (see also Fig. 3): some trees are almost flat, others may have branches up to 5 levels deep. The identification of the trees of a forest is characterized by the presence of various types of possible errors:

1. at the single node level:
   (a) inability to assign a node to any tree (the data does not provide evidence high enough to allow the attribution);
   (b) attribution of a node to a branch to which it does not belong (even if the structure of the tree has been detected correctly);
   (c) attribution of a node to a tree to which it does not belong.
2. at the "branch" level (node groups):
   (a) correct identification of the individual branches, but no final connection between the various parts of the tree;
   (b) branches composed of heterogeneous parts (whole branches of the same tree, or subsets of other branches of the same tree);
   (c) fusion between two branches of two different trees.

In the list above, the errors are presented in increasing order of seriousness: the non-attribution of a node to a tree is less dangerous than its attribution to an incorrect branch, which in turn is less dangerous than its attribution to an incorrect tree. In the same way, the correct identification of a branch permits its subsequent analysis using cancer evolution inference methods, which is not possible in the case of the union of branches belonging to different trees.

Interestingly, in absence of noise the RI methodology never inferred erroneous attributions of nodes, while the fraction of unassigned nodes was always extremely low (a fact leading to an average node coverage close to 0.99 - see Fig. 2).

The trees have always been correctly identified, while the main branches of those classified as "not completely identified" have always been identified: indeed, when the identification is not complete only the final fusion of the two already identified branches is missing. This phenomenon mainly occurs when it is necessary to identify large trees, as shown also in Fig. 2, where the number of complete identifications increases with the number of trees (because of the constant number of nodes in each experiment, the average size of the trees decreases with their number). Indeed, in order to be statistically significant, the complete identification of large objects requires a high number of observations (a number that depends on the structure of the tree, and that is sometimes greater than the number of available patients).[1]

---

[1] Note that this observation is related to the number of observations that is possible to have in currently available clinical studies, rather than to the method we are applying.

Size of trees					Noise 0.0			Noise 0.1				Noise 0.25			
Tree 4	Tree 3	Tree 2	Tree 1	Configuration	Incorrect attributions	Unassigned nodes	Trees / total trees	Incorrect attributions	Unassigned nodes	Trees / total trees	Branch merge	Incorrect attributions	Unassigned nodes	Trees / total trees	Branch merge
			3	17	F2T1	0	2	1/2	0	0	1/2	0	0	1	1/2	0
			1	19	F2T2	0	2	1/2	0	3	1/2	0	1	1	0/2	1
			1	19	F2T3	0	1	1/2	0	2	1/2	0	0	1	0/2	0
			3	17	F2T4	0	0	2/2	0	1	1/2	0	0	1	1/2	0
			8	12	F2T5	0	0	2/2	0	0	0/2	0	0	0	0/2	0
		1	6	13	F3T1	0	0	3/3	0	1	2/3	0	0	1	2/3	0
		1	3	16	F3T2	0	2	2/3	1	2	1/3	0	0	2	2/3	0
		2	3	15	F3T3	0	2	2/3	0	2	2/3	0	0	2	2/3	0
		2	4	14	F3T4	0	1	2/3	0	2	2/3	0	0	2	2/3	0
		1	7	12	F3T5	0	0	2/3	1	0	1/3	1	1	2	1/3	1
1	2	7	10	F4T1	0	0	4/4	0	1	1/4	0	1	1	0/4	1
1	2	7	10	F4T2	0	0	4/4	0	0	3/4	0	1	2	2/4	1
1	1	8	10	F4T3	0	1	3/4	0	1	2/4	0	1	2	1/4	1
3	4	4	9	F4T4	0	0	4/4	0	1	2/4	0	0	2	2/4	1
1	1	4	14	F4T5	0	0	3/4	0	0	3/4	0	0	1	2/4	0

Fig. 2. Summary of the results of the analysis. For each configuration "FxTy" (where x stands for "number of trees in the forest" and y for "case identifier"), some relevant quantities are shown: on the right, the results of the analysis for different noise levels; on the left, the size of the trees. In case of absence of noise, the number of incorrect assignments, the number of unassigned nodes and the number of trees found compared to the number of total trees are shown. In the presence of noise, the number of incorrect fusions of different branches is also present (there are no incorrect fusions of branches in absence of noise). Note that an incorrect merger causes the inclusion of a tree consisting of a single element in a branch of a different tree (a situation corresponding to the only incorrect node assignment - see the text for a more detailed comment).

Anyhow, even in case of big progression trees, the branches are always correctly identified: therefore, this information might be employed by a cancer evolution inference method without many concerns (a branch is actually a coherent tree). This allows us to say that the information resulting from the analysis of the noiseless cases can provide the information needed for a correct downstream processing.

Similar considerations can be made even when the noise level is considerably increased (Fig. 2). The number of node assignment errors remains extremely low, and the "merging" of different branches (which occurred only once with noise 0.1) are actually fusions of a corrected branch with a single element belonging to another branch. Indeed, these events are almost exclusively trees composed of a single node, which in these cases are melted in a branch of a larger tree. Indeed, because of the presence of noise, the occurrence of a single mutation becomes the occurrence of more-than-one mutations: this causes the presence of a tree composed of a single element to be implausible[2]. Moreover, these mergers take place at very low zI values: always below 5.0, and mostly just near the 3.0 threshold.

---

[2] The identification of a tree composed of a single element is a case that is strongly influenced by noise.

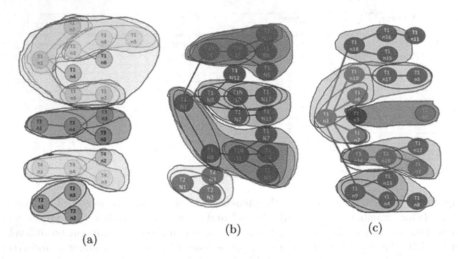

(a)   (b)   (c)

Fig. 3. Results of the analysis on simulated cross-sectional cancer datasets sampled from forest topologies. The figures show all the groups gradually identified, up to the large final groups: the circles identify the nodes of each tree, connected by evolutionary paths. The fact that each small group resulting from the analysis is completely included in a larger group is a consequence of the use of the iterated sieve algorithm. (a) A perfect identification - F4T4 configuration without noise. (b) Even if the limited number of patients does not allow the identification of a very large tree, the RI algorithm still manages to identify the main branches of each tree; in this case there is no evidence which allows one to attribute the T1n12 node to one tree rather than to the other one - F2T1 configuration, two trees (one completely identified by the yellow set), with noise 0.25. (c) The F2T2 configuration is composed by two trees, respectively of size 1 and 19. When the noise level is equal to 0.25, one of the last groupings (at a very low evidence level) fuses a rare leaf of the largest tree with the single element of the second tree. Node T1n11 is not assigned to any tree.

5   Conclusions

The capacity of detecting the presence of groups of variables that appear to be well coordinated among themselves and have a weaker interaction with the remainder of the system can often lead to a high-level description of the dynamical organization of a complex system, and thus to its understanding.

In this work we extend an interesting kind of analysis based on the juxtaposition of dynamical states belonging to different instantiations of a complex system, to a particular case in which different patients suffering from the same disease can be considered special "instances" of the same pathological dynamic process. In order to assess the accuracy of the approach, we simulated 200 patients, of which we therefore knew the division into classes. The RI methodology, based on entropic measurements, allowed us to identify distinct cancer subtypes (represented in this paper by different trees, each including successive accumulating mutations), potentially a very useful information for subsequent analyses or treatments. Vice versa - and in a very interesting way - this analysis confirmed the possibility of inferring information on the dynamic organization of a process starting from the observation of its distinct instances, under the simple hypothesis of a common structural condition. These results support further investigations, in two main directions. On one side, we plan to identify suitable data from real cases that can be used to test our approach in ordinary situations. On the other side, we plan to increase the theoretical bases of the RI methodology and deepen the concepts related to the dynamic organization of complex systems.

Acknowledgments. The work of Laura Sani was supported by a grant from Regione Emilia Romagna ("Creazione di valore per imprese e società con la gestione e l'analisi di BIG DATA", POR FSE 2014/2020, Obiettivo Tematico 10) within the "Piano triennale alte competenze per la ricerca, il trasferimento tecnologico e l'imprenditorialità" framework, and by Infor srl.

Marco Villani thanks the support provided by the FAR2019 project of the Department of Physics, Informatics and Mathematics of the University of Modena and Reggio Emilia.

References

1. Bar-Yam, Y.: Dynamics of Complex Systems. Studies in Nonlinearity. Perseus Publishing, Reading (1997)
2. Bazzi, M., Porter, M.A., Williams, S., McDonald, M., Fenn, D.J., Howison, S.D.: Community detection in temporal multilayer networks, with an application to correlation networks. Multiscale Model. Simul. **14**(1), 1–41 (2016)
3. Beerenwinkel, N., Schwarz, R.F., Gerstung, M., Markowetz, F.: Cancer evolution: mathematical models and computational inference. Syst. Biol. **64**(1), e1–e25 (2014)
4. Bennett, J.M., Catovsky, D., Daniel, M.T., Flandrin, G., Galton, D.A., Gralnick, H.R., Sultan, C.: Proposals for the classification of the acute leukaemias french-american-british (fab) co-operative group. Br. J. Haematol. **33**(4), 451–458 (1976)
5. Burrell, R.A., McGranahan, N., Bartek, J., Swanton, C.: The causes and consequences of genetic heterogeneity in cancer evolution. Nature **501**(7467), 338–345 (2013)
6. Caravagna, G., Graudenzi, A., Ramazzotti, D., Sanz-Pamplona, R., De Sano, L., Mauri, G., Moreno, V., Antoniotti, M., Mishra, B.: Algorithmic methods to infer the evolutionary trajectories in cancer progression. Proc. Natl. Acad. Sci. **113**(28), E4025–E4034 (2016)
7. Cover, T., Thomas, A.: Elements of Information Theory, 2nd edn. Wiley-Interscience, New York (2006)
8. Davis, A., Gao, R., Navin, N.: Tumor evolution: linear, branching, neutral or punctuated? Biochim. Biophys. Acta (BBA) Rev. Cancer **1867**(2), 151–161 (2017)
9. Gao, Y., Church, G.: Improving molecular cancer class discovery through sparse non-negative matrix factorization. Bioinformatics **21**(21), 3970–3975 (2005)
10. Gillies, R.J., Verduzco, D., Gatenby, R.A.: Evolutionary dynamics of carcinogenesis and why targeted therapy does not work. Nat. Rev. Cancer **12**(7), 487 (2012)

11. Hofree, M., Shen, J.P., Carter, H., Gross, A., Ideker, T.: Network-based stratification of tumor mutations. Nat. Methods **10**(11), 1108 (2013)
12. Holme, P., Saramäki, J.: Temporal networks. Phys. Rep. **519**(3), 97–125 (2012)
13. Hordijk, W., Steel, M.: Detecting autocatalytic, self-sustaining sets in chemical reaction systems. J. Theor. Biol. **227**(4), 451–461 (2004)
14. Lane, D., Pumain, D., van der Leeuw, S.E., West, G.: Complexity Perspectives in Innovation and Social Change, vol. 7. Springer Science & Business Media, Dordrecht (2009)
15. Loohuis, L.O., Caravagna, G., Graudenzi, A., Ramazzotti, D., Mauri, G., Antoniotti, M., Mishra, B.: Inferring tree causal models of cancer progression with probability raising. PLoS ONE **9**(10), e108358 (2014)
16. Lu, J., et al.: Microrna expression profiles classify human cancers. Nature **435**(7043), 834 (2005)
17. Merlo, L.M., Pepper, J.W., Reid, B.J., Maley, C.C.: Cancer as an evolutionary and ecological process. Nat. Rev. Cancer **6**(12), 924 (2006)
18. Network, C.G.A., et al.: Comprehensive molecular characterization of human colon and rectal cancer. Nature **487**(7407), 330 (2012)
19. Nik-Zainal, S., Van Loo, P., Wedge, D.C., Alexandrov, L.B., Greenman, C.D., Lau, K.W., Raine, K., Jones, D., Marshall, J., Ramakrishna, M., et al.: The life history of 21 breast cancers. Cell **149**(5), 994–1007 (2012)
20. Nowell, P.C.: The clonal evolution of tumor cell populations. Science **194**(4260), 23–28 (1976)
21. Papoulis, A., Pillai, S.U.: Probability, Random Variables, and Stochastic Processes. McGraw-Hill, Boston (2015)
22. Ramazzotti, D., Caravagna, G., Olde Loohuis, L., Graudenzi, A., Korsunsky, I., Mauri, G., Antoniotti, M., Mishra, B.: Capri: efficient inference of cancer progression models from cross-sectional data. Bioinformatics **31**(18), 3016–3026 (2015)
23. Ramazzotti, D., Graudenzi, A., De Sano, L., Antoniotti, M., Caravagna, G.: Learning mutational graphs of individual tumour evolution from single-cell and multi-region sequencing data. BMC Bioinform. **20**(1), 210 (2019)
24. Righi, R., Roli, A., Russo, M., Serra, R., Villani, M.: New paths for the application of DCI in social sciences: theoretical issues regarding an empirical analysis. In: Rossi, F., Piotto, S., Concilio, S. (eds.) WIVACE 2016. CCIS, vol. 708, pp. 42–52. Springer, Cham (2017). https://doi.org/10.1007/978-3-319-57711-1_4
25. Roli, A., Villani, M., Caprari, R., Serra, R.: Identifying critical states through the relevance index. Entropy **19**(2), 73 (2017)
26. Sani, L., Amoretti, M., Vicari, E., Mordonini, M., Pecori, R., Roli, A., Villani, M., Cagnoni, S., Serra, R.: Efficient search of relevant structures in complex systems. In: Adorni, G., Cagnoni, S., Gori, M., Maratea, M. (eds.) AI*IA 2016. LNCS (LNAI), vol. 10037, pp. 35–48. Springer, Cham (2016). https://doi.org/10.1007/978-3-319-49130-1_4
27. Sani, L., D'Addese, G., Pecori, R., Mordonini, M., Villani, M., Cagnoni, S.: An integration-based approach to pattern clustering and classification. In: Ghidini, C., Magnini, B., Passerini, A., Traverso, P. (eds.) AI*IA 2018. LNCS (LNAI), vol. 11298, pp. 362–374. Springer, Cham (2018). https://doi.org/10.1007/978-3-030-03840-3_27
28. Sani, L., Lombardo, G., Pecori, R., Fornacciari, P., Mordonini, M., Cagnoni, S.: Social relevance index for studying communities in a Facebook group of patients. In: Sim, K., Kaufmann, P. (eds.) EvoApplications 2018. LNCS, vol. 10784, pp. 125–140. Springer, Cham (2018). https://doi.org/10.1007/978-3-319-77538-8_10

29. Silvestri, G., Sani, L., Amoretti, M., Pecori, R., Vicari, E., Mordonini, M., Cagnoni, S.: Searching relevant variable subsets in complex systems using K-Means PSO. In: Pelillo, M., Poli, I., Roli, A., Serra, R., Slanzi, D., Villani, M. (eds.) WIVACE 2017. CCIS, vol. 830, pp. 308–321. Springer, Cham (2018). https://doi.org/10.1007/978-3-319-78658-2_23

30. Suppes, P.: A probabilistic theory of causality (1973)

31. Tononi, G., McIntosh, A., Russel, D., Edelman, G.: Functional clustering: identifying strongly interactive brain regions in neuroimaging data. Neuroimage **7**, 133–149 (1998)

32. Tononi, G., Sporns, O., Edelman, G.M.: A measure for brain complexity: relating functional segregation and integration in the nervous system. Proc. Natl. Acad. Sci. **91**(11), 5033–5037 (1994)

33. Vicari, E., Amoretti, M., Sani, L., Mordonini, M., Pecori, R., Roli, A., Villani, M., Cagnoni, S., Serra, R.: GPU-based parallel search of relevant variable sets in complex systems. In: Rossi, F., Piotto, S., Concilio, S. (eds.) WIVACE 2016. CCIS, vol. 708, pp. 14–25. Springer, Cham (2017). https://doi.org/10.1007/978-3-319-57711-1_2

34. Villani, M., Filisetti, A., Benedettini, S., Roli, A., Lane, D., Serra, R.: The detection of intermediate-level emergent structures and patterns. In: Miglino, O., et al. (eds.) Advances in Artificial Life, ECAL 2013, pp. 372–378. The MIT Press (2013). http://mitpress.mit.edu/books/advances-artificial-life-ecal-2013

35. Villani, M., Roli, A., Filisetti, A., Fiorucci, M., Poli, I., Serra, R.: The search for candidate relevant subsets of variables in complex systems. Artif. Life **21**(4), 412–431 (2015)

36. Villani, M., Sani, L., Amoretti, M., Vicari, E., Pecori, R., Mordonini, M., Cagnoni, S., Serra, R.: A relevance index method to infer global properties of biological networks. In: Pelillo, M., Poli, I., Roli, A., Serra, R., Slanzi, D., Villani, M. (eds.) WIVACE 2017. CCIS, vol. 830, pp. 129–141. Springer, Cham (2018). https://doi.org/10.1007/978-3-319-78658-2_10

37. Villani, M., Sani, L., Pecori, R., Amoretti, M., Roli, A., Mordonini, M., Serra, R., Cagnoni, S.: An iterative information-theoretic approach to the detection of structures in complex systems. Complexity **2018**, 1–15 (2018). https://doi.org/10.1155/2018/3687839

38. Vogelstein, B., Papadopoulos, N., Velculescu, V.E., Zhou, S., Diaz, L.A., Kinzler, K.W.: Cancer genome landscapes. Science **339**(6127), 1546–1558 (2013)

The Simulation of Noise Impact on the Dynamics of a Discrete Chaotic Map

Uladzislau Sychou[✉][iD]

Laboratory of Robotic Systems, United Institute of Informatics Problems
of National Academy of Sciences of Belarus, Minsk, Belarus
vsychyov@robotics.by

Abstract. There is no clear understanding of the way how a noise influence on nonlinear dynamical systems with chaotic behaviour. Due to difficulties with analysis of real-world chaotic oscillators computer simulation is preferable. However, during the process of simulation, a noise presents too. This kind of noise is called the quantization noise and it leads to appear of limit cycles in simulations. It is customary to avoid such limit cycles. However, it is important to understand how it arises and behaves. This work presents brief results related to the simulation of the logistic map under a noise impact.

Keywords: Chaos · Chua's circuit · Quantization noise · Limit cycle

1 Introduction

The effect of noise on chaotic systems has been investigated in the number of researches [1,2]. Some of them study a noise impact on the dynamics of nonlinear systems. Other ones are aimed to adequately approximate noisy systems. Since computers and data acquisition devices have become widely used in studies, a noise impact should be dealing with more carefully due to quantization noise [3].

This noise is commonly considered as an effect that can be avoided by means of appropriate data types. The reason to avoid it is that the quantization noise leads to limit cycles in simulations. This effect also calls "digital degradation". Taking into account that chaotic systems haven't period at all [4], this effect is considered as an undesirable effect because it restricts the long-term simulations of chaotic dynamics. However, in most scientific papers the exact properties of quantization noise haven't taken into account, being limited only common thoughts. As a result, it can be seen that the double-precision data type accepted as sufficient for most calculations. It is true for common scientific calculations but not always suitable for chaotic systems. The high sensitivity of chaotic systems leads to the fact that the quantization noise has a great impact on system dynamics. So, it is highly desirable to collect experimental results about a noise impact.

Supported by the Swiss State Secretariat for Education, Research and Innovation grant No. SFG 450 and grant BRFFI-RFFI No. F18R-229.

2   The Computer Simulation

Given the above, it can be proposed to use the actual quantization noise as the only type of noise acts in the simulation to obtain pure data about a noise impact. And the easiest way to accurately set up the noise level is to use the fixed-point arithmetic so that the length of the registers $Qm.n$ allows to specify the noise level by varying the fraction length n in the range from 6 to 30 bits. This research has been conducted using MATLAB fixed-point toolbox on the Logistic map:

$$x_i = a - x_{i-1}^2 \qquad (1)$$

In x_i-notation is used instead of x_n-notation, which is common for discrete systems, to avoid confusion because the symbol n is used above as part of the Q-notation for fixed-point options. During the simulations rounding to the nearest has been used. Overflows have been handled using the saturation method.

For the logistic map Largest Lyapunov Exponent has been calculated using Rosenstein's Algorithm [5]. The value of the exponent is 0.6607. So the map can be considered as chaotic.

The signal-to-noise ratio (SNR) can be calculated using the formula:

$$SNR = 6.02Q + 10.8 - 20log_{10}\frac{X_m}{\sigma_x} \qquad (2)$$

where Q - word length, X_m - full-scale value, σ_x - root mean square of a map. The fixed-point arithmetic as a noise source allows preventing variables from going beyond the limits on which the functions are defined. It would have happened if the floating-point arithmetic had been used. As already said, the quantization noise produces a period p during simulation of the chaotic map $f : X \to X$:

$$x_{j+p} = x_j \qquad (3)$$

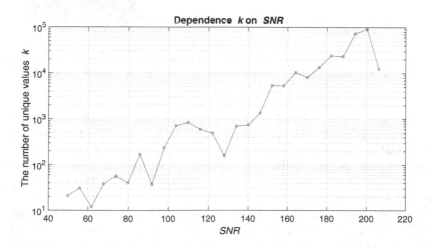

Fig. 1. The dependence of a number k of unique values of x on SNR.

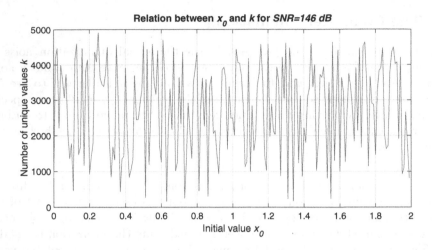

Fig. 2. The number of unique values k depends on an initial condition x_0.

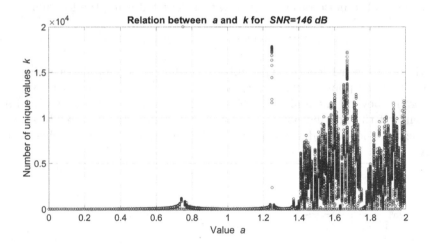

Fig. 3. The influence of a parameter a on a number of unique values of x.

where $x \in X$, $j = 1, 2, 3....$ There are cases when equation becomes true after some step [6]. In the such maps value x_i, $i = 1, 2, 3...$ depends on x_{i-1} only. It makes it possible to estimate the size of the period using the number of unique values of x. For one-dimensional maps condition $k < 2^Q$ is true, where k - number of unique values of x and $k = j + p$. Below it assumed that $k \approx p$, so exactly k is used to estimate periodization.

The first task of the simulation is to estimate how the value k depends on the SNR (see Fig. 1). It can be seen that the graph fluctuated, although SNR grows exponentially. Such behaviour can be explained using the dependence of vakue k on an initial condition x_0 (see Fig. 2). The graph in Fig. 2 shows that the value k depends on the initial value x_0. Moreover, not only x_0 sets a length of a period but value a too. The graph in Fig. 3 shows the result of the simulation when a has been changed from 0 to 1.99. Each circle in each column represents the value k obtained for each x_0, while x_0 has been changing for each value of a from $x_0 = 0$ to $x_0 = 1.99$. The plot in Fig. 3 shows that the number of unique values depends on initial conditions only in the chaotic regime, which is observed from $a \approx 1.5$ to $a = 1.99$ except a little window near the $a \approx 1.8$.

3   Conclusions

Due to sensitivity to initial conditions the dynamics of the fixed-point simulation of the logistic map can be considered as chaotic. The observed periodization has a number of appropriate features. A period length depends on the signal-to-noise ratio of a system. In the chaotic mode, there is the nonlinear dependence on initial conditions.

References

1. He, T., Habib, S.: Chaos and noise. Chaos: Interdiscip. J. Nonlinear Sci. **23**(3) (2013). https://doi.org/10.1063/1.4813864
2. Gyebrószki, G., Csernák, G.: Structures within the quantization Noise: micro-chaos in digitally controlled systems. In: Part of special issue: 12th IFAC Symposium on Robot Control SYROCO 2018. Elsiver (2018). https://doi.org/10.1016/j.ifacol.2018.11.551
3. Widrow, B., Kollár, I.: Quantization Noise. Cambridge University Press, Cambridge (2008). https://doi.org/10.1017/CBO9780511754661
4. Moon, F.: Chaotic Vibrations. Wiley, New York (2004)
5. Rosenstein, M.T., et al.: A practical method for calculating largest Lyapunov exponents from small data sets. Physica D **65**, 117–134 (1993). https://doi.org/10.1016/0167-2789(93)90009-P
6. Crownover, R.M.: Introduction to Fractals and Chaos. Jones and Bartlett Publishers Inc., Sudbury (1995)

Exploiting Distributed Discrete-Event Simulation Techniques for Parallel Execution of Cellular Automata

Andrea Giordano[1], Donato D'Ambrosio[2], Alessio De Rango[3],
Alessio Portaro[2], William Spataro[2(✉)], and Rocco Rongo[2]

[1] ICAR-CNR, Rende, Cosenza, Italy
giordano@icar.cnr.it
[2] Department of Mathematics and Computer Science,
University of Calabria, Rende, Italy
{d.dambrosio,spataro,rongo}@unical.it
[3] Department of Environmental Engineering,
University of Calabria, Rende, Italy
alessio.derango@unical.it

Abstract. The Cellular Automata (CA) paradigm is well-suited to model complex systems based on local rules of evolution such as those related to fluid-dynamics, crowd simulation, fire propagation and many more. In addition, CA can be profitably exploited as a support for different kinds of numerical approaches, such as finite element and finite volume methods. As the size of the problem increases, a cellular automaton can be easily parallelized through domain partitioning in order to scale up its execution. However, the performance and scalability of cellular automata executed on parallel/distributed machines are limited by the necessity of synchronizing all the nodes at each time step, i.e., a node can execute a new step only after all the other nodes have executed the previous one. This paper presents a preliminary study on how techniques taken from the Discrete-Event Simulation field can be adopted for the parallelization of CA on distributed memory architectures with the goal of reducing the synchronization burden. In particular, we combine the active/inactive cells technique, which is well-known in the CA context, with the concept of lookahead which, instead, is adopted in the field of distributed discrete-event simulation research.

Keywords: Cellular Automata · Discrete-Event Simulation · Lookahead

1 Introduction

Parallel Computing methodologies are adopted for the efficient solutions of large data-intensive computational problems. The development of High Performance Computing (HPC) permits the adoption of numerical simulations as a tool for

© Springer Nature Switzerland AG 2020
F. Cicirelli et al. (Eds.): WIVACE 2019, CCIS 1200, pp. 66–77, 2020.
https://doi.org/10.1007/978-3-030-45016-8_8

solving complex equation systems which govern the dynamics of real complex phenomena as fire spreading [29], unmanned aerial vehicles simulation [5], lava flow [28] or swarm intelligence algorithms [3].

An important categorization of parallel computing architectures concerns their memory layout, where machines are divided in: (i) *shared memory architectures*, where all the processing elements share a common memory (e.g., in the case of a multi-core computing system) and (ii) *distributed memory architectures (DMA)*, where each processing element has its own memory and communication among each of them occurs via remote operations. A classical parallel programming model for distributed memory architectures is MPI (Message Passing Interface) [20]. MPI is mainly used to send/receive data among processes in an optimized fashion, and can be suitably adopted also in the case of shared memory architectures. Nevertheless, OpenMP [2] is the best choice in the latter kind of memory layout since it offers a set of API functions dedicated to the shared memory programming model.

Numerical methods, such as Cellular Automata (CA) and the Finite Volume Method (FEM), are widely adopted for the simulation of natural phenomena. In particular, despite their simple definition, CA [30] may give rise to extremely complex behavior at a macroscopic level and are widely adopted for the simulation of complex natural processes. In fact, even if local laws that determine the dynamics of the system are known, the system's global behavior can be hard to be predicted, giving rise to what is called *emergent behaviour*. CA applications and models have been developed by numerous researchers, and regard both theoretical and scientific fields (e.g., [1,7,10,12,13,15,21–26,31]). Thanks to their parallel nature, CA models can be efficiently parallelized on a set of computing nodes to speed up their execution [14]. In particular, the CA space is partitioned by adopting a typical data-parallel scheme in regions, where the execution of each of them is assigned to a different computing node.

In this paper, we present the application of techniques derived from the Discrete Event Simulation (DES) field with the aim of improving the parallel execution for CA when implemented in a distributed memory environment. In particular, we adopt the lookahead technique [17], originally applied for mitigating synchronization issues in DES, combined with the *active* cells optimization, which is well-known in the CA context (e.g. [8,27]). The paper is organized as follows: Sect. 2 reports a quick overview of CA parallel partitioning and optimization techniques; Sect. 3 presents the adopted DDES technique with the aim in improving execution times; finally, Sect. 4 concludes the paper with a general discussion about the presented research and outlooks for possible future work.

2   Parallel Execution of Cellular Automata in Distributed Memory Architectures

Cellular Automata (CA) is a parallel computational paradigm which can fruitfully be adopted in treating complex systems whose behaviour may be expressed in terms of local laws. CA were introduced by von Neumann in the 1950s to study self-reproduction issues, and are particularly appropriate to model complex systems that are characterized by an elevated number of interacting elementary components. By applying relatively simple rules, the global behaviour of the system emerges from the local interactions of its elementary (cellular) units. Moreover, from a computational point of view, they are equivalent to Turing Machines [6].

A CA can be intuitively thought as a d-dimensional space (the cellular space), partitioned into cells of uniform size, representing a finite automaton (f_a). Input for each f_a is given by the state of the cell and the state of the cell's neighbourhood. This is defined by means of a geometrical pattern, which is invariant in time and constant over the cells. At time $t = 0$, corresponding to step 0, the f_a's states define the CA initial configuration. At following steps, the CA evolves by means of the f_a's *transition function*, which is simultaneously applied to each f_a.

CA execution on computers is straightforward and concerns the evaluation of the transition function over the cellular space, which is evaluated sequentially or in a parallel fashion. At step t, since a cell transition function evaluation requires the state of neighbouring cells at step $t - 1$, these latter must be kept in memory during the execution, giving rise to what is called "maximum parallelism" behaviour. This issue can be simply addressed by adopting two matrices for the space state: the *read* and the *write* matrix. In particular, during the execution of a generic step, the transition functions are evaluated by reading states from the read matrix and by writing the results to the write matrix. Subsequently, i.e., after all transition functions have been evaluated, the two matrices are swapped and the next computational step can begin.

Due to their inherent parallel nature, CA execution can be efficiently parallelized [11]. In particular, the approach of partitioning the space (or region) and assigning each subspace to a specific processing element is a good strategy to improve the computational performances of the system. A mono-dimensional or a two-dimensional partitioning [4,19] schema is usually adopted when partitioning a cellular automata (Fig. 1). Each partition of the space is assigned to a different computing node, which is in charge of applying the transition function to all the cells belonging to that specific partitioning.

Since the computation of a transition function of a cell is based on the states of the cell's neighbourhood, these can overlap more regions in case of cells located at the edge of a region, as seen in Fig. 2. As a consequence, the execution of the transition function for these cells requires information that is located in adjacent computing nodes. Thus, at each computing step, the states of border cells (or *halo* cells) need to be exchanged among adjacent nodes in order to keep the parallel execution consistent. In turn, each border area of a region is composed of two distinct parts: the *local border* and the *mirror border*. Figure 3 shows the borders in the cases of mono-dimensional space partitioning. As seen, the local border is managed by the local node and its content replicated in the mirror border of the adjacent nodes.

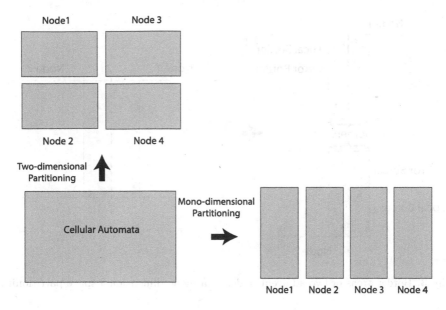

Fig. 1. The cellular space partitioned into regions which are associated with parallel computing nodes. Two alternative types of partitioning are shown, mono-dimensional and two-dimensional.

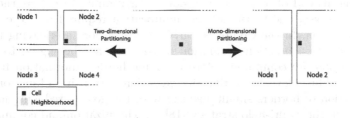

Fig. 2. The neighbourhood of a edge cell overlapping more regions

2.1 Optimization of CA Implementations

While a simple data-parallel partitioning alone can already improve computational performances in CA execution, several other optimizations can further reduce execution times. A first and widely adopted strategy consists in the hyper-rectangular bounding box (HRBB) optimization [9]. In many CA applications, the system's dynamics only affects a small region of the whole computational domain, as for the case of topologically connected phenomena, like debris or lava flows. In these cases, a standard approach where the overall domain is processed can lead to a considerable waste of computational resources. Therefore, the HRBB approach consists in surrounding the simulated phenomenon by means of a fitting rectangle (in case of a 2D model), by restricting the computation to this specific sub-region. A similar approach is adopted by the *quantization* optimization (e.g., cf. [10]), which consists in adopting a dynamic set of coordinates only

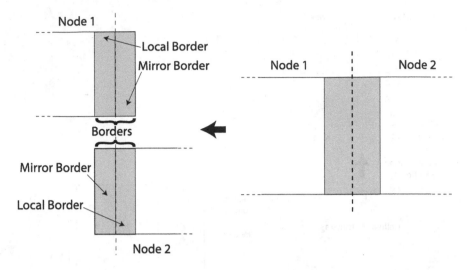

Fig. 3. Border areas of two adjacent nodes in a mono-dimensional space partitioning

of the active cells during the simulation, by restricting the computation only to this set. Authors in [4] have shown the advantage of using the two-dimensional partitioning instead of a mono-dimensional partitioning in CA as the number of nodes increases, due to the less communication required in the exchanging borders operation. In the Computation/Communication interleaving technique [18], the transition function evaluation of the cells not belonging to the edge part of a region are computed before receiving border information from adjacent regions. This allows to overlap the execution of transition function and the communication of borders simultaneously with the goal of reducing idle times. Eventually, in the Multi-halo strategy [18], synchronization and communication burden is reduced by decreasing the number of messages needed between adjacent nodes, by sending few large messages (i.e., more border regions) rather than many small messages. In particular, the strategy introduces a sort of redundancy in computation with the aim of reducing the synchronization points required for exchanging borders.

3   Distributed Discrete Event Simulation

The Discrete-Event Simulation paradigm deals with the modeling of the evolution of a real system described by a typically irregular sequence of consecutive (discrete) states. In general, the evolution of a system's state stays unvaried for a definite amount of time which precedes an instant when it evolves *immediately*, entering a new state. Based on when the change of system state occurs, we can devise two kinds of evolution: (*i*) an Event-driven execution, where an event internal or external to the simulation causes a change to the system's state and the simulated time, or (*ii*) a time-stepped execution, where the evolution evolves step by step. As evident, a time stepped approach seems appropriate for modeling a CA, since its evolution consists in reading the state at the step t of the Simulated Time and then computing the state of the CA at the time $t + 1$, without any other requirement.

3.1   The Synchronization Problem and Lookahead

A Distributed (or Parallel) Discrete-event Simulation (DDES) [17] can be viewed as a collection of sequential discrete event simulations that interact by exchanging timestamped messages. Each message represents an event that is scheduled (sent) by one simulator to another. At a generic simulation instant, it could be possible that the *wallclock* time of two different processes have reached different points in the *simulated* time. In particular, let's call these two processes LP1 (*Logical Process* 1) and LP2, and let's say LP1 has reached a point into the simulated time successive to the point reached from LP2. For the LP1 and LP2 respective computations, the discrepancy between the simulated time instants could represent a hitch, if LP2 affects LP1 in its past. LP1 in fact would need to handle that event which potentially could have changed its state if it had arrived earlier in the wallclock time. In this case, LP1 is said to violate the *Local Causality Constraint* (LCC) [16], defined in literature as a simulation consisting in logical processes where each LP manages events in non-decreasing timestamp order. If this is the case, i.e., the LCC is satisfied, then the outcome of a DDES will be exactly that of the corresponding (sequential) DES [16]. To obviate the violation of the LCC, two main approaches are adopted: (*i*) A *Conservative Approach*, in which an LP evolves only if it has the assurance in not receiving any message or event that could affect its past states and a (*ii*) *Non Conservative* (or *Optimistic*) approach, where a LP "hopes" to not receive an event which may affect its past and evolves normally. In this latter case, however, if an event is received, it has to perform a so called roll-back operation. As evident, though introducing a major overhead, an optimistic approach exploits a greater degree of parallelism w.r.t. conservative approaches. For instance, if two events affect each other over time but not "so frequent", optimistic mechanisms can process the events concurrently, while conservative methods must inevitably introduce a serialization of the execution.

Among conservative approaches to solve the LCC problem, the lookahead approach [16] consists in a LP "promising" not to produce an output event before a given time. Specifically, if a LP at simulation time T can only schedule new events with timestamp of at least $T + L$, then L is referred to as the *lookahead* for the logical process. By exploiting this property, it is possible to reduce the synchronization burden of the parallel system thus achieving improvements of performances, as detailed in the following section.

3.2   Application of DES Techniques in Parallel Execution of CA

As evident in the CA context, synchronization among different nodes (i.e., each LP) is mandatory at each computational step in order to maintain consistency.

The aim of the proposed research is to *relax* the synchronization burden required in the CA paradigm in a parallel and distributed environment, by exploiting the lookahead technique.

For illustrative purposes, let us focus on a not infrequent CA evolution situation, where an LP contains a border that has not changed its state after the application of the transition function in a certain step. In this case, no communication to an adjacent LP is necessary, with the possibility in achieving an increase in performances. However, if this optimization is exploited *as is*, a possible deadlock situation could occur since the *neighbor* LP expects border communication at the same step of the evolution. Thus, a mechanism to allow for every LP to know when to receive or send a border is needed. In other words, each LP requires a *lookahead* information for its neighbours. In order to devise such information, in this work we exploit the states of *idle* (or stationary) cells, defined as those cells whose state does not change after the application of the transition function[1]. Let us now focus on the evolution of a cell with a neighbouring idle cell. For instance, it might be possible that under certain conditions, such cell could take two steps to change its state. By applying the same reasoning to cells that are more and more distant from the considered one, we can assume that the number of steps required to change their state increases. In this way, each LP can find a suitable value of lookahead for its borders by simply considering the number idle cells in its own cellular space and in its adjacent neighbours. As a consequence, the lookahead which is considered has to be directly proportional to the distance from the nearest active cell of each border.

Figure 4 shows an example of the dynamics of a CA. Here, an initial set of active cell located on a LP are at a distance of 6 cells from the right border, progressively evolving towards right after 8 CA steps to a configuration which is 2 cells from the edge part of the partitioning. In this case, a lookahead value of 6 (i.e., the minimum distance in cells from active cells to the border) can be considered, which represents the minimum number of CA steps that the CA can evolve without performing border exchange for synchronization. As evident, this results in a communication burden decrease with the consequence of improving computational performances.

The module pseudocode which manages the lookahead implemented in addition to a normal CA execution loop is described in Algorithm 1. While the stopping criterion is not met (line 1), the lookahead module manages three distinct phases: (i) the LP sending phase (from line 3), (ii) the LP receiving phase (from line 14) and (iii) the normal CA execution and new lookahead evaluation (from line 25). During the LP sending phase (i.e., when an LP is involved in sending a communication to a neighbour LP), the lookahead value for each LP is decremented by one if it is greater than zero (lines 5–7), meaning that no border exchange takes place as the lookahead phase is still active. However, when the value reaches zero (line 8), the new lookahead value is computed (line 9) and checked if it is zero (line 10), meaning border send must take place (line 11) or,

---

[1] For instance, for a debris flow CA model, an idle cell is a cell which does not contain landslide matter.

Algorithm 1: Lookahead module pseudocode

```
   // LA = Lookahead
   // LP = Logical Process
1  while !StopCriterion()                              // CA stop criterion met?
2  do
3     forall LP sending                                // LP sending phase
4     do
5        if lastLookahead > 0                          // still in LA phase
6        then
7           | lookahead = lookahead - 1       // decrease last lookahead sent
8        else
9           newLookahead = ComputeNewLookahead()          // compute new
            lookahead value
10          if newLookahead = 0 then
11             | SendNewBorder()                       // Send border info
12          else
13             | SendNewLookahead()            // Send new lookahead only

14    forall LP receiving                              // LP receving phase
15    do
16       if LastLookahead > 0                          // still in LA phase
17       then
18          | lookahead = lookahead - 1       // decrease last lookahead sent
19       else
20          waitNeighbourNotification()       // wait info from neighbour
21          if notificationIsBorder() then
22             | updateCellularSpaceBorder()       // received border info
23          else
24             updateLastReceivedLookaheadVal()           // received new
               lookahead only

25    forall cells ∈ SpacePortion do
26       transitionFunction()
27       if cellIdle() then
28          forall Border ∈ SpacePortion do
29             if cellIsNearestBorder() then
30                | computeNewLookahead()

31    step ← step + 1                                  // Next CA step
```

---

more favorably, the simulation can advance for another lookahead steps which are sent to neighbour LPs (line 13). The LP receiving phase (i.e., when an LP is involved in receiving a communication from a neighbour LP) management takes place from line 14. Here, after a control verifying if the evolution is in the lookahead phase (i.e., no border swap is occurring) at line 16, the decrement of the

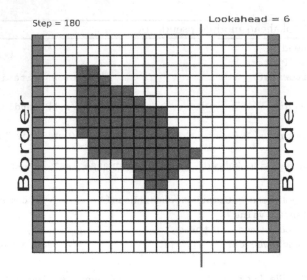

(a) At step 180 the minimum distance of an active cell (in red) to the right border is 6, which is assumed as the lookahead value (i.e., steps to be evolved without border exchange)

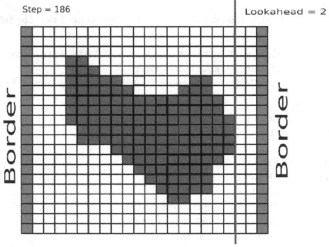

(b) After the 6 computational steps, the minimum distance of an active cell to the right border is now 2, which can be assumed as the new value of lookahead for the next CA evolution.

Fig. 4. Example of CA evolution of a region with lookahead, assuming expansion of the dynamics rightwards. (Color figure online)

lookahead value is done at line 18. Otherwise, some message is awaited from a neighbour LP (line 20) which can represent either a border (at line 22) or a new lookahead value (at line 24). Eventually, the third phase regarding normal CA execution takes place from line 25. Here, after the application of the transition function, the new lookahead value computation is carried out consisting, in this preliminary work, as simply the *distance* of the nearest idle cell to the border.

4 Conclusions

This paper presents the adoption of a conservative strategy adopted in DDES (Distributed Discrete Event Simulation) literature for speeding up the parallel execution of CA. Specifically, the algorithm reduces the synchronization burden which is required at each computational step by adopting the lookahead technique, consisting in the (minimum) number of CA steps that a LP carry our without requiring neighbour information of neighbour nodes.

Future work will regard the application of the proposed technique for optimizing real CA models. Experiments will also be carried out on different node partitionings, such as two-dimensional, or by coping the proposed technique with other CA optimizations found in literature, such as multi-halo and/or communication/computation interleaving techniques (cf. [18]).

Eventually, future developments will regard the application of the non-conservative (or optimistic) mechanisms with the aim of hopefully leading to better timings under favorable conditions. In fact, this technique consists in avoiding for the borders exchange with all the neighboring nodes, *optimistically* applying the transition function to the node's cells, and performing a roll back operation whenever a consistency error is detected, in order to communicate the missing information that corrupted the execution.

References

1. Aidun, C., Clausen, J.: Lattice-Boltzmann method for complex flows. Ann. Rev. Fluid Mech. **42**, 439–472 (2010)
2. Chapman, B., Jost, G., Van Der Pas, R.: Using OpenMP: Portable Shared Memory Parallel Programming, vol. 10. MIT Press, Cambridge (2008)
3. Cicirelli, F., Forestiero, A., Giordano, A., Mastroianni, C.: Transparent and efficient parallelization of swarm algorithms. ACM Trans. Auton. Adapt. Syst. (TAAS) **11**(2), 14 (2016)
4. Cicirelli, F., Forestiero, A., Giordano, A., Mastroianni, C.: Parallelization of space-aware applications: modeling and performance analysis. J. Netw. Comput. Appl. **122**, 115–127 (2018)
5. Cicirelli, F., Furfaro, A., Giordano, A., Nigro, L.: An agent infrastructure for distributed simulations over HLA and a case study using unmanned aerial vehicles. In: 40th Annual Simulation Symposium, 2007, ANSS 2007, pp. 231–238. IEEE (2007)
6. Cook, M.: Universality in elementary cellular automata. Complex Syst. **15**(1), 1–40 (2004)

7. Crisci, G.M., Gregorio, S.D., Rongo, R., Spataro, W.: PYR: a cellular automata model for pyroclastic flows and application to the 1991 Mt. Pinatubo eruption. Future Gener. Comput. Syst. **21**(7), 1019–1032 (2005)
8. D'Ambrosio, D., et al.: The open computing abstraction layer for parallel complex systems modeling on many-core systems. J. Parallel Distrib. Comput. **121**, 53–70 (2018)
9. D'Ambrosio, D., Filippone, G., Marocco, D., Rongo, R., Spataro, W.: Efficient application of GPGPU for lava flow hazard mapping. J. Supercomput. **65**(2), 630–644 (2013)
10. Di Gregorio, S., Filippone, G., Spataro, W., Trunfio, G.: Accelerating wildfire susceptibility mapping through GPGPU. J. Parallel Distrib. Comput. **73**(8), 1183–1194 (2013)
11. D'Ambrosio, D., Filippone, G., Rongo, R., Spataro, W., Trunfio, G.A.: Cellular automata and GPGPU: an application to lava flow modeling. Int. J. Grid High Perform. Comput. **4**(3), 30–47 (2012)
12. Filippone, G., D'Ambrosio, D., Marocco, D., Spataro, W.: Morphological coevolution for fluid dynamical-related risk mitigation. ACM Trans. Model. Comput. Simul. (TOMACS) **26**(3), 18 (2016)
13. Folino, G.: Cellar: a high level cellular programming language with regions, pp. 259–266 (February 2000)
14. Folino, G.: Predictability of cellular programs implemented with camelot, pp. 468–474 (February 2001)
15. Frish, U., Hasslacher, B., Pomeau, Y.: Lattice gas automata for the Navier- Stokes equation. Phys. Rev. Lett. **56**(14), 1505–1508 (1986)
16. Fujimoto, R.: Parallel and distributed simulation systems. In: Proceeding of the 2001 Winter Simulation Conference (Cat. No.01CH37304), vol. 1, pp. 147–157 (2000)
17. Fujimoto, R.M.: Parallel and Distribution Simulation Systems, 1st edn. Wiley, New York (1999)
18. Giordano, A., De Rango, A., D'Ambrosio, D., Rongo, R., Spataro, W.: Strategies for parallel execution of cellular automata in distributed memory architectures. In: 2019 27th Euromicro International Conference on Parallel, Distributed and Network-Based Processing (PDP), pp. 406–413. IEEE (2019)
19. Giordano, A., et al.: Parallel execution of cellular automata through space partitioning: the landslide simulation sciddicas3-hex case study. In: 2017 25th Euromicro International Conference on Parallel, Distributed and Network-based Processing (PDP), pp. 505–510. IEEE (2017)
20. Gropp, W.D., Gropp, W., Lusk, E., Skjellum, A., Lusk, A.D.F.E.E.: Using MPI: portable parallel programming with the message-passing interface, vol. 1 (1999)
21. Higuera, F., Jimenez, J.: Boltzmann approach to lattice gas simulations. Europhys. Lett. **9**(7), 663–668 (1989)
22. Langton, C.: Computation at the edge of caos: phase transition and emergent computation. Phys. D **42**, 12–37 (1990)
23. LucÀ, F., D'Ambrosio, D., Robustelli, G., Rongo, R., Spataro, W.: Integrating geomorphology, statistic and numerical simulations for landslide invasion hazard scenarios mapping: an example in the sorrento peninsula (Italy). Comput. Geosci. **67**(1811), 163–172 (2014)
24. McNamara, G., Zanetti, G.: Use of the Boltzmann equation to simulate lattice-gas automata. Phys. Rev. Lett. **61**, 2332–2335 (1988)
25. Ninagawa, S.: Dynamics of universal computation and 1/f noise in elementary cellular automata. Chaos Solitons Fractals **70**(1), 42–48 (2015)

26. Ntinas, V.G., Moutafis, B.E., Trunfio, G.A., Sirakoulis, G.C.: Parallel fuzzy cellular automata for data-driven simulation of wildfire spreading. J. Comput. Sci. **21**, 469–485 (2017)
27. Rango, A.D., Spataro, D., Spataro, W., D'Ambrosio, D.: A first multi-GPU/multi-node implementation of the open computing abstraction layer. J. Comput. Sci. **32**, 115–124 (2019)
28. Spataro, D., D'Ambrosio, D., Filippone, G., Rongo, R., Spataro, W., Marocco, D.: The new SCIARA-fv3 numerical model and acceleration by GPGPU strategies. Int. J. High Perform. Comput. Appl. **31**(2), 163–176 (2017)
29. Trunfio, G.A., D'Ambrosio, D., Rongo, R., Spataro, W., Di Gregorio, S.: A new algorithm for simulating wildfire spread through cellular automata. ACM Trans. Model. Comput. Simul. (TOMACS) **22**(1), 6 (2011)
30. Von Neumann, J., Burks, A.W., et al.: Theory of self-reproducing automata. IEEE Trans. Neural Netw. **5**(1), 3–14 (1966)
31. Wolfram, S.: A New Kind of Science. Wolfram Media Inc., Champaign (2002)

A Relevance Index-Based Method for Improved Detection of Malicious Users in Social Networks

Laura Sani[1(✉)], Riccardo Pecori[1,2,3(✉)], Paolo Fornacciari[1],
Monica Mordonini[1], Michele Tomaiuolo[1], and Stefano Cagnoni[1(✉)]

[1] Dip. di Ingegneria e Architettura, Università di Parma, Parma, Italy
{laura.sani,stefano.cagnoni}@unipr.it
[2] Dip. di Ingegneria, Università del Sannio, Benevento, Italy
rpecori@unisannio.it
[3] SMARTEST Research Centre, Università eCampus, Novedrate, CO, Italy

Abstract. The phenomenon of "trolling" in social networks is becoming a very serious threat to the online presence of people and companies, since it may affect ordinary people, public profiles of brands, as well as popular characters. In this paper, we present a novel method to preprocess the temporal data describing the activity of possible troll profiles on Twitter, with the aim of improving automatic troll detection. The method is based on the zI, a Relevance Index metric usually employed in the identification of relevant variable subsets in complex systems. In this case, the zI is used to group different user profiles, detecting different behavioral patterns for standard users and trolls. The comparison of the results, obtained on data preprocessed using this novel method and on the original dataset, demonstrates that the technique generally improves the classification performance of troll detection, virtually independently of the classifier that is used.

Keywords: Complex systems · Social networks · Troll detection · Relevance index

1 Introduction

The definition of an "Internet troll" is still debated, since the acknowledgement of the associated behaviors is largely subjective. This lack of a single and precise definition has made the phenomenon vague and, consequently, has arisen little interest within the researchers' community. Nevertheless, it has very serious implications for the online presence of both ordinary people and public profiles of brands, as well as popular characters. As a matter of fact, quite recently, this problem has received much more attention by the general public, largely as a response to events that have caused a stir about the "toxicity" of some social media sites [10]. Moreover, popular magazines and specialized press agencies have begun addressing the issue as well as writing widely-read articles on such a topic.

F. Cicirelli et al. (Eds.): WIVACE 2019, CCIS 1200, pp. 78–89, 2020.
https://doi.org/10.1007/978-3-030-45016-8_9

A troll can be defined as an individual adopting an antisocial behavior, who provokes and irritates other users of an online social platform with the intended effect to derail the normal evolution of an online discussion and possibly to stop it abruptly. To this aim, he/she can adopt an aggressive or offensive language, with prolonged arguments and unsolicited personal attacks. Thus, the term "trolling" refers to "a specific type of malicious online behavior, intended to disrupt interactions, aggravate interactional partners and lure them into fruitless argumentation" [2].

To address this problem, online platforms like Twitter periodically release new services and features, such as the functionalities that allow one to report anti-social behaviors or to "mute" annoying users. However, all attempts made till now have not been able to eradicate the problem, while the need to deal with it is becoming more and more urgent. Hence, it is fundamental to create automated methods to manage the complexity of the issue, possibly leveraging on artificial intelligence, data mining and social-network analysis.

Therefore, some studies have been conducted to demonstrate the applicability of various analytical tools and methods. For example, in [6], a set of objective features extracted from social network logs has been identified as characterizing malicious trolling behaviors. Among these, features related with time and frequency of actions appear to be the most relevant for troll detection.

This work aims at deepening the analysis of the impact of such temporal features on troll detection. The final goal of our study is to improve the classification accuracy achieved by the time-related features by themselves, by clustering training data and substituting them with the centroid of their cluster. Clustering is performed using a method, based on the Relevance Index (RI) metrics [22], for preprocessing the data that have been used as the training set from which a troll detector has been derived in [6].

In [12], we demonstrated that the Relevance Index metrics, which were originally developed for the analysis of the structure of complex systems, could also be used effectively as a data clustering tool. While that paper was mainly a proof of concept, in this work we evaluate the RI approach on a practical case which we show can benefit from its application.

The rest of the manuscript is organized as follows: the next section shortly introduces the state of the art of the research on troll detection along with past usages of the RI metrics; then we describe the zI method and present a zI-based preprocessing technique which improves the classification accuracy of troll detection; finally, some concluding remarks seal up the article.

2   Related Work

Although the concept of Internet trolling may seem to be tied to concepts like rudeness, arrogance, impertinence and vulgarity, the latter do not provide an accurate description of the phenomenon since, typically, trolling also implies keeping the real intent of causing problems hidden. It is clear that trolling is a more complex problem than just provocative attacks. Accordingly, the task of automatic troll detection needs to take many different aspects into account [6].

Among the different viewpoints from which the phenomenon can be studied, some researches show that temporal patterns of the online behavior can provide a valuable contribution to automatic troll detection. Various studies have related the frequency of publication of posts and comments to the quality of online discussions. The behavior of users who would be later banned from some large websites has been analyzed in [1], highlighting the features that characterize such behaviors. In addition to the kind of text produced by users before being banned, the paper also observes patterns of activity. In particular, among the most useful features for detecting users likely to be banned in the future, an important role is played by the frequency of some activities, like the number of posts and comments per day. In [4], the authors use newsroom interviews, reader surveys and moderators' choices for characterizing the comments published on a newspaper website. In particular, they note that the frequency of commenting is a valuable indicator of low-quality discourse.

In [8], two classifiers are described, which detect two types of trolls: "paid trolls", who try to manipulate a user's opinion, and "mentioned trolls", who offend users and provoke anger. Many features regarding sentiment and text analysis are considered, based either on lexicon models or on bag-of-words models. Moreover, some metadata are considered, including the publication time. In particular, the analysis distinguishes a worktime period and a nighttime period; it also distinguishes workdays (Monday-Friday) and weekend days (Saturday and Sunday). Quite interestingly, the analysis demonstrates that these features provide the largest impact on accuracy.

Troll Pacifier, the holistic system proposed in [6], includes features which represent a breakdown of the daily activity into four-hour long time windows. This time window interval has been chosen after an extensive comparison, in which the authors have trained automatic classifiers based on different algorithms (K-Nearest Neighbors (KNN), Naïve Bayes (NB), Sequential Minimal Optimization (SMO), J48, etc.) using different time window lengths. Features measuring the activities within four-hour long intervals provided consistently the best classification results. In TrollPacifier, the time intervals represent single days (from Sunday to Saturday). In addition to these metrics, additional features consider the frequency of actions in the recent timeline and during the whole time of a user's presence on the platform.

As regards the method we use in this paper, the Relevance Index was originally introduced in [5, 21] as the *Dynamical Cluster Index* (DCI), an extension of the Cluster Index (CI) defined by Tononi [17,18] in the '90s. More recently, the DCI has been applied to different scenarios to detect relevant subsets of variables in several complex systems, such as abstract models of gene regulatory networks and simulated chemical [20] and biological systems [22], as well as social networks [13]. This last work was a first attempt to apply the index to a social network scenario, trying to reveal communities and subsets of users within a specific Facebook group.

Along the years, the DCI has been gradually refined and paired with a proper metaheuristics [11] and a GPU implementation [19], in order to face its intrinsically high computational complexity. During this evolution, the DCI has been termed Relevance Index (RI), while the iterative application of a sieve to the results of the RI computation has allowed researchers to remove redundant variable sets, as well as to merge the original variables according to an optimal sequence, obtaining a hierarchical functional representation of the overall complex system [23].

More recently, the RI has become a family of indexes (RI metrics) which have been also used for purposes different from the detection of subsets in complex systems. In particular, one of these indexes, the so-called zI, based on information integration, has been employed for feature extraction [14, 15], as well as for pattern clustering and classification [12], with promising preliminary results.

3   Method

In this section, we describe the theoretical framework within which one can define the zI index and its computation using a metaheuristic and an iterative sieving procedure. Moreover, we also explain the variable-grouping action of the zI and its usage for troll detection, which will lead to the results described in the following section.

3.1   The zI Index

A system U, described by N random variables (RVs), can be characterized by its state $\bar{X}_U = [X_1, ..., X_N]$, i.e., a random vector whose elements are the N random variables which, within the context described in this paper, represent the users. Given this scenario, we consider n independent observations of the state $\{\bar{X}_U(i)\}_{i=1}^n$. Therefore, for each RV X_j, we have a sequence $\{X_j(1), .., X_j(n)\}$ of independent random variables, identically distributed (iid). If we consider n to be sufficiently large, we have that, from the Asymptotic Equipartition Property (AEP) [3], the empirical entropy of a sequence of iid RVs, computed through the relative frequencies, converges in probability (ip) to the real individual entropy.

We have referred to entropies because they are involved in the definition of the *integration*, in turn at the basis of the computation of the zI. The integration measures the mutual dependence among the k elements in S_k and is defined as the difference between the summation of the entropies of the single variables composing subset S_k and the total entropy of subset S_k itself:

$$I(S_k) = \sum_{j=1}^{k} H(X_j) - H(X^k) \tag{1}$$

where $H(X^k)$ is the entropy [3] of the sequence $\{X_1, ..., X_k\}$. It can be observed that $I(S_k) \geq 0$ and it is zero when the RVs are independent, such that $H(X^k) = \sum_{j=1}^{k} H(X_j|X^{j-1}) = \sum_{j=1}^{k} H(X_j)$. On the other hand, the integration increases with the decrease of the second term, i.e., with the increase of the correlation among the random variables.

Based on the integration, the zI metric can be defined as follows:

$$zI(S_k) = \frac{2nI(S_k) - \langle 2nI(S_{NULL}) \rangle}{\sigma(2nI(S_{NULL}))} \tag{2}$$

where n is the number of observations or instances of the whole system, S_k a subset of k out of N variables and S_{NULL} is a subset of dimension k extracted from a null system U_h (i.e., a system within which all N variables are independent).

It can be demonstrated that

$$2nI \approx \chi_g^2 \tag{3}$$

under the null hypothesis of independent RVs and with n large enough.

Moreover, it can be demonstrated that the degrees of freedom of the Chi-square distribution of Eq. 3 can be expressed as:

$$g = (\prod_{j=1}^{N} |T_j| - 1) - (\sum_{j=1}^{N} (|T_j| - 1)) \tag{4}$$

where $|T_j|$ is the cardinality of the alphabet of X_j. The second term of Eq. 4 is the number of degrees of freedom under the null hypothesis, while the first term is the number of degrees of freedom in the opposite case, in which no RV describing the system is independent of the others.

It is known [9] that the mean and the standard deviation of the chi-squared distribution with g degrees of freedom are

$$\mu = g \tag{5}$$
$$\sigma = \sqrt{2g} \tag{6}$$

Hence, the elements of Eq. 2 are all defined and can be computed easily.

3.2   zI Computation

The zI expresses the relevance of a group of variables for the system under consideration: the higher the zI, the more relevant the group. A list of relevant sets could be obtained in principle by enumerating all possible subsets of the system variables and ranking them according to the zI values. However, this exhaustive search is not feasible for systems described by a large number of variables, because the number of subsets increases exponentially with the latter. When large systems are analyzed, this issue makes it impossible to compute the zI for every possible subset of variables, even using massively parallel hardware such as GPUs.

Therefore, in order to quickly find the most relevant subsets and to efficiently face the complexity of the exhaustive search, we used a metaheuristic (HyReSS), which hybridizes a genetic algorithm with local search strategies, using the same configuration as in [11,16]. HyReSS maximizes, as its fitness function, the zI expression reported in Eq. 2. The search procedure is guided by the statistics, computed at runtime, on the results that the algorithm itself is obtaining. HyReSS searches the N_s highest-zI sets by exploring many peaks in parallel; this happens because the evolutionary search is enhanced through a niching technique, which tries to maintain population diversity. The genetic algorithm on which HyReSS is based is run first, to address the search towards the basins of attraction of the main local maxima in the search space. Then, the regions identified during the evolutionary process are explored, more finely and extensively, by means of a series of local searches, in order to further improve the results.

The zI computation, which is the most computation-intensive module within the algorithm, is parallelized for large blocks of sets by means of a CUDA[1] kernel which fits the computational needs of this problem particularly well [19].

As a further and final step, a sieving algorithm [5], performed iteratively, is used to reduce the list of N_s sets found by HyReSS to the most representative ones. The filtering performed by the sieving algorithm is based on a reasonable criterion: if set S_1 is a proper subset, or superset, of set S_2 and ranks higher than S_2 according to the zI, then S_1 is considered more relevant than S_2. Therefore, the algorithm keeps only those sets that are not included in, or do not include, any other higher-zI set. The sieve stops in case no more eliminations are possible, i.e., the remaining subsets can not be decomposed any further.

The sieve allows one to analyze the organization of the variables in terms of the lowest-level subsets. In order to deepen the analysis, considering also the aggregated hierarchical relations among the identified sets, we employed an *iterative version* of the sieve, acting on the data by iteratively grouping one or more sets into a single entity. This is performed by iteratively running the sieving algorithm on the same data, each time using a new representation of the variables, where the top-ranked set, in terms of zI value, of the previous iteration is considered atomic and is replaced by a single variable (group variable) [23]. In this way, each iteration produces a new representation made of both single variables and of the group variables that have been previously detected.

3.3   zI-Based Clustering

As described in Sect. 3.1, the zI method relies on the analysis of a set of observations which can be represented as a matrix where each row represents the status of the system at a certain time, while each column represents a status variable. Analyzing these observations, the zI method is able to find groups of variables (columns) that exhibit correlated properties or behaviors.

In [12], we demonstrated that the zI method, originally developed for complex system analysis, can also be used effectively as a pattern clustering tool. In that case, the columns of the data matrix represent the patterns while the rows represent the features by which each pattern is described. Thus, when the zI is

---

[1] https://developer.nvidia.com.

computed, the iterative sieving algorithm operates by subsequently merging the highest-zI set of patterns into a new cluster in each step.

In the following section, we present a practical case in which we apply the zI-based clustering as a preprocessing step for troll detection.

3.4   zI-Based Preprocessing for Troll Detection

The problem has been approached in a supervised way, applying the iterative sieving algorithm and computing the zI index from data regarding trolls and legitimate users separately, to derive two specific models of the two classes. Thus, we analyzed two dynamical systems, including 300 trolls and 300 standard users, respectively. For each system, the zI analysis is intrinsically unsupervised and aims at detecting different behavioral patterns of the same class of users, finding clusters of trolls and clusters of non-trolls characterized by a similar dynamical behavior.

Each observation (data matrix row) represents the frequency of the tweets posted by each user (data matrix column) within a specific time interval during the day (e.g., the frequency of user's tweets posted on Monday from 00:00 to 04:00). The extremes of the four-hour long ranges have been chosen in order to create non-overlapping intervals distinguishing worktime and nighttime periods, but also working days and weekend days.

Given the large number of variables which describe the two systems, we used HyReSS combined with the iterative sieving algorithm, as described in Sect. 3.2, to identify the most relevant groups of trolls and legitimate users. Each group of trolls (or legitimate users) corresponds to a behavioral prototype. Based on this assumption, we created a new preprocessed dataset composed of the centroids of the groups of trolls and legitimate users discovered by our method. As will be shown in Sect. 4, this preprocessing phase allowed us to improve the classification accuracy in the automatic troll detection task.

The main building blocks of our method are shown in Fig. 1, which also illustrates the data flow of the procedure by which we compared the classifications obtained from the raw data and those obtained from the zI-preprocessed ones.

4   Experimental Results

In order to apply the zI method to the analysis of the dynamical behavior of trolls and legitimate users on Twitter, we used the labeled dataset described in [6]. In this section, we describe the dataset we preprocessed using the zI approach and show the accuracy improvement that could be achieved using the new dataset thus obtained to train different classifiers.

4.1   Dataset Description

The dataset described in [6] has been obtained using two cascaded approaches. The first one is based on distant supervision [7] and allows one to obtain a

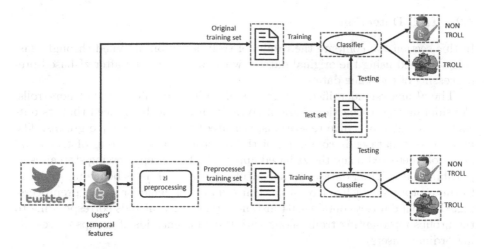

Fig. 1. Troll detection on Twitter using the original dataset and the zI-preprocessed dataset as training sets for automatic classification.

raw dataset. The second one consists in manually filtering the previous dataset in order to improve the dataset reliability. In particular, users reported to the official support channels have been classified as "trolls" after manually inspecting both their recent timelines and their role and attitude in prolonged discussions, where they were repeatedly mentioned as "trolls".

The dataset contains six groups of features corresponding to six approaches to troll detection, each of which relies on different kinds of information (sentiment analysis, time and frequency of actions, text content and style, user behavior, interactions with the community, advertisement of external content). In this work we focus on the group of features related to time and frequency of actions, after we observed in [6] that this set of features can give a significant contribution to troll detection. In particular, we perform a dynamical analysis of the frequency of users' posts, to improve classification accuracy. As stated in Sect. 2, we measured the considered features according to four-hour long intervals, starting at 00:00 daily (i.e., Sunday from 00:00 to 04:00, Sunday from 04:00 to 08:00, Sunday from 08:00 to 12:00, ..., Saturday from 20:00 to 00:00) since these intervals provided consistently the best classification results in an extensive comparison campaign carried out in [6]. In our experiments, we considered a four-level discretization of the original frequency values (low, medium-low, medium-high and high frequency of users' posts in each time interval).

The dataset comprises data about 500 trolls and 500 non-trolls. The instances have been split into a training set composed of 300 trolls and 300 non-trolls and a test set composed of the remaining 200 trolls and 200 non-trolls.

4.2 Troll Detection

In this section, we compare the results of troll detection obtained through classifiers trained using the original dataset with those obtained after zI-based preprocessing of the same data.

The zI analysis identified 161 groups of trolls and 143 groups of non-trolls. Considering the stochastic nature of HyReSS, it should be noticed that its outcome was stable over five repetitions, repeatedly finding the same groups. The preprocessed dataset is composed of the centroids of each group of trolls and non-trolls detected using the zI-based approach. This preprocessing step aims at polishing the original dataset by substituting groups of users behaving similarly to the prototype which is closest to their dynamical behavior. After preprocessing, the final dataset contains 304 instances (the 161 centroids corresponding to the groups representing trolls, along with the 143 centroids of groups representing ordinary users).

The effect of this preprocessing on the detection of troll and non-troll users has been tested using four different classifiers: Random Forest (RF), Naïve Bayes (NB), Sequential Minimal Optimization (SMO), and K-Nearest Neighbors (KNN). Table 1 shows the accuracies obtained by training the classifiers on the original and on the preprocessed dataset.

To take into consideration the intrinsic stochastic nature of Random Forest and Sequential Minimal Optimization, we repeated all experiments using such classifiers five times; in the table, we report the average accuracy over the five runs.

Table 1. Troll detection accuracies obtained by training different classifiers on the original dataset and on the preprocessed one.

Dataset	RF	NB	SMO	KNN (k = 1)	KNN (k = 5)	KNN (k = 10)
Original	76.95%	60.00%	65.00%	65.75%	70.75%	70.75%
Preprocessed	77.10%	63.25%	69.25%	69.25%	71.25%	72.75%

As shown in Table 1, the introduction of the zI-based preprocessing allowed us to improve the accuracy of all the classifiers.

Since the preprocessed dataset is not perfectly balanced (161 trolls and 143 non-trolls corresponding to the centroids of the troll and non-troll groups), in Table 2 we also show the F-measure of the classifiers for a fairer comparison.

The classifiers trained after the zI-based preprocessing tend to perform better than the standard ones. The accuracy and F-measure improvements highlight the importance of the dynamical analysis of users' behavior for automatic troll detection.

Table 2. F-measure obtained by training different classifiers on the original dataset and on the preprocessed one.

Dataset	RF	NB	SMO	KNN (k = 1)	KNN (k = 5)	KNN (k = 10)
Original	0.769	0.546	0.622	0.656	0.704	0.704
Preprocessed	0.771	0.594	0.678	0.689	0.709	0.725

5 Conclusion

In this work, we applied a dynamical analysis based on the zI metric to preprocess the temporal data describing the users' behavior, aimed at improving the classification accuracy of troll detection. The dynamical analysis we propose is based on the observation of the frequency of users' tweets within different time windows.

We compared the results of troll detection based on the raw dataset previously used with the same goal with those obtained after zI-based preprocessing of the same data. The tests have been performed using four different classifiers.

The classifiers trained using the zI-based preprocessing performed better than the standard ones, leading to an improvement in the classification accuracy and F-measure. The results show that the zI analysis is able to detect significant behavioral patterns for troll detection, finding clusters of trolls and clusters of non-trolls characterized by a similar dynamical behavior.

As future work, we aim at using the zI to analyze also other features relevant to the detection of trolls.

Acknowledgments. The work of Laura Sani was supported by a grant from Regione Emilia Romagna ("Creazione di valore per imprese e società" con la gestione e l'analisi di BIG DATA", POR FSE 2014/2020, Obiettivo Tematico 10) within the "Piano triennale alte competenze per la ricerca, il trasferimento tecnologico e l'imprenditorialità" framework, and by Infor srl.

References

1. Cheng, J., Danescu-Niculescu-Mizil, C., Leskovec, J.: Antisocial behavior in online discussion communities. arXiv preprint arXiv:1504.00680 (2015)
2. Coles, B.A., West, M.: Trolling the trolls: online forum users constructions of the nature and properties of trolling. Comput. Hum. Behav. **60**, 233–244 (2016)
3. Cover, T.M., Thomas, J.A.: Elements of Information Theory (Wiley Series in Telecommunications and Signal Processing). Wiley-Interscience, New York (2006)
4. Diakopoulos, N., Naaman, M.: Towards quality discourse in online news comments. In: Proceedings of the ACM 2011 Conference on Computer Supported Cooperative Work, pp. 133–142. ACM (2011)
5. Filisetti, A., Villani, M., Roli, A., Fiorucci, M., Serra, R.: Exploring the organisation of complex systems through the dynamical interactions among their relevant subsets. In: Andrews, P., et al. (eds.) Proceedings of the European Conference on Artificial Life 2015, ECAL 2015, pp. 286–293. The MIT Press (2015)

6. Fornacciari, P., Mordonini, M., Poggi, A., Sani, L., Tomaiuolo, M.: A holistic system for troll detection on Twitter. Comput. Hum. Behav. **89**, 258–268 (2018)
7. Go, A., Bhayani, R., Huang, L.: Twitter sentiment classification using distant supervision. CS224N Project Report, Stanford **1**(12), 1–6 (2009)
8. Mihaylov, T., Nakov, P.: Hunting for troll comments in news community forums. In: Proceedings of the 54th Annual Meeting of the Association for Computational Linguistics, vol. 2, pp. 399–405 (2016)
9. Papoulis, A., Pillai, S.U.: Probability, Random Variables, and Stochastic Processes. McGraw-Hill, Boston (2015)
10. Rosenbaum, S.: Is Twitter toxic? Can social media be tamed? Forbes 2016 (Sep 9) (2016)
11. Sani, L., et al.: Efficient search of relevant structures in complex systems. In: Adorni, G., Cagnoni, S., Gori, M., Maratea, M. (eds.) AI*IA 2016. LNCS (LNAI), vol. 10037, pp. 35–48. Springer, Cham (2016). https://doi.org/10.1007/978-3-319-49130-1_4
12. Sani, L., D'Addese, G., Pecori, R., Mordonini, M., Villani, M., Cagnoni, S.: An integration-based approach to pattern clustering and classification. In: Ghidini, C., Magnini, B., Passerini, A., Traverso, P. (eds.) AI*IA 2018. LNCS (LNAI), vol. 11298, pp. 362–374. Springer, Cham (2018). https://doi.org/10.1007/978-3-030-03840-3_27
13. Sani, L., Lombardo, G., Pecori, R., Fornacciari, P., Mordonini, M., Cagnoni, S.: Social relevance index for studying communities in a facebook group of patients. In: Sim, K., Kaufmann, P. (eds.) EvoApplications 2018. LNCS, vol. 10784, pp. 125–140. Springer, Cham (2018). https://doi.org/10.1007/978-3-319-77538-8_10
14. Sani, L., Pecori, R., Mordonini, M., Cagnoni, S.: From complex system analysis to feature extraction: a novel unsupervised feature extraction method based on the relevance index metrics. Computation **7**, 39 (2019)
15. Sani, L., Pecori, R., Vicari, E., Amoretti, M., Mordonini, M., Cagnoni, S.: Can the relevance index be used to evolve relevant feature sets? In: Sim, K., Kaufmann, P. (eds.) EvoApplications 2018. LNCS, vol. 10784, pp. 472–479. Springer, Cham (2018). https://doi.org/10.1007/978-3-319-77538-8_32
16. Silvestri, G., et al.: Searching relevant variable subsets in complex systems using k-means PSO. In: Pelillo, M., Poli, I., Roli, A., Serra, R., Slanzi, D., Villani, M. (eds.) WIVACE 2017. CCIS, vol. 830, pp. 308–321. Springer, Cham (2018). https://doi.org/10.1007/978-3-319-78658-2_23
17. Tononi, G., McIntosh, A., Russel, D., Edelman, G.: Functional clustering: identifying strongly interactive brain regions in neuroimaging data. Neuroimage **7**, 133–149 (1998)
18. Tononi, G., Sporns, O., Edelman, G.M.: A measure for brain complexity: relating functional segregation and integration in the nervous system. Proc. Natl. Acad. Sci. **91**(11), 5033–5037 (1994)
19. Vicari, E., et al.: GPU-based parallel search of relevant variable sets in complex systems. In: Rossi, F., Piotto, S., Concilio, S. (eds.) WIVACE 2016. CCIS, vol. 708, pp. 14–25. Springer, Cham (2017). https://doi.org/10.1007/978-3-319-57711-1_2
20. Villani, M., Filisetti, A., Benedettini, S., Roli, A., Lane, D., Serra, R.: The detection of intermediate-level emergent structures and patterns. In: Miglino, O., et al. (eds.) Advances in Artificial Life, ECAL 2013. pp. 372–378. The MIT Press (2013)
21. Villani, M., Roli, A., Filisetti, A., Fiorucci, M., Poli, I., Serra, R.: The search for candidate relevant subsets of variables in complex systems. Artificial Life **21**(4) (2015)

22. Villani, M., et al.: A relevance index method to infer global properties of biological networks. In: Pelillo, M., Poli, I., Roli, A., Serra, R., Slanzi, D., Villani, M. (eds.) WIVACE 2017. CCIS, vol. 830, pp. 129–141. Springer, Cham (2018). https://doi.org/10.1007/978-3-319-78658-2_10
23. Villani, M., et al.: An iterative information-theoretic approach to the detection of structures in complex systems. Complexity **2018**, 1–15 (2018)

An Analysis of Cooperative Coevolutionary Differential Evolution as Neural Networks Optimizer

Marco Baioletti, Gabriele Di Bari, and Valentina Poggioni[✉]

Dip. Matematica e Informatica, Università di Perugia, Perugia, Italy
{marco.baioletti,valentina.poggioni}@unipg.it,
gabriele.dibari@unifi.it

Abstract. Differential Evolution for Neural Networks (DENN) is an optimizer for neural network weights based on Differential Evolution. Although DENN has shown good performance with middle-size networks, the number of weights is an evident limitation of the approach. The aim of this work is to figure out if coevolutionary strategies implemented on top of DENN could be of help during the optimization phase. Moreover, we studied two of the classical problems connected to the application of evolutionary computation, i.e. the stagnation and the lack of population diversity, and the use of a crowding strategy to address them. The system has been tested on classical benchmark classification problems and experimental results are presented and discussed.

Keywords: Neuroevolution · Differential Evolution · Coevolution

1 Introduction

Differential Evolution (DE) is a well known evolutionary algorithm and its behaviour as weight optimizer for neural networks has been studied in several works, as described in [4]. Despite some results about the stagnation problems [10], there are recent interesting works showing the efficiency of DE, and in general EA approaches, to neural network weights optimization [2,3,8,12,14,16].

In this paper we focus on the DENN algorithm [3], which is a DE algorithm for neural network optimization and which is able to solve classification problems with a large number of parameters. DENN proposes a direct encoding of neural networks into population elements, i.e each element contains the connection weights and the bias of the corresponding network. It implements several DE versions with different crossover and mutation operators. The objective of the evolution is to find the network that minimize the cross entropy function. In order to reduce the computational cost of the fitness evaluation, DENN employs a batch system similar to the limited evaluation technique proposed by [8]. Although DENN has shown good performance also on networks of moderate size, the number of weights is an evident limitation of this approach. This is the main reason that led us to test the application of co-evolution

© Springer Nature Switzerland AG 2020
F. Cicirelli et al. (Eds.): WIVACE 2019, CCIS 1200, pp. 90–99, 2020.
https://doi.org/10.1007/978-3-030-45016-8_10

strategies and to implement the system CoDE, with the idea to divide the effort into different subpopulations that can evolve separately. In particular we tested two different coevolutive strategies: CoDE-Layer, that implements the coevolution at the network layer level (a subpopulation for each network layer), and CoDE-Matrix, that implements the coevolution at the network component level (a subpopulation for each weights and bias matrix).

By analyzing the accuracy results obtained in preliminary experiments, we found a problem with population diversity and we decided to implement the crowding strategy [13] in order to maintain the population diversity and hence to increase the possibility to find a better solution. The experiments over some classification benchmark problems aim to study the impact of using the batch system, coevolution and crowding on DENN.

2   Background

2.1   Differential Evolution

Differential Evolution (DE) is an evolutionary algorithm used for optimization over continuous spaces, which operates by improving a population of N candidate solutions evaluated by means of a fitness function f though a iterative process. The first phase is the initialization, in which the first population is (randomly or accordingly to a given strategy) generated. Then, for each generation a new population is computed though mutation and crossover operators; each new vector is evaluated and then the best ones are chosen, according to a selection operator, for the next generation. The evolution may proceed for a fixed number of generations or until a given criterion is met.

The mutation used in DE is called *differential mutation*. For each vector *target vector* x_i, for $i = 1, \ldots, N$, of the current generation, a vector $\bar{y}_i$, called *donor vector*, is obtained as linear combination of some vectors in the population selected according to a given strategy [4,5]. There exist many variants of the mutation operator (see for instance [4]). In this work we have implemented and used the current_to_pbest (1) mutation operator, that is defined as

$$\bar{y}_i = x_i + F(x_{pbest} - x_i) + F(x_a - x_b) \tag{1}$$

where $p \in (0, 1]$ and pbest is randomly selected index from the indices of the best $N \times p$ elements of the population. Moreover, x_b is an individual randomly chosen from the set

$$\{x_1, \ldots, x_N\} \setminus \{x_a, x_i\} \cup \mathcal{A}$$

where $\mathcal{A}$ is an external archive of bounded size (usually with at most N elements) that contains the individuals discarded by the selection operator.

The crossover operator creates a new vector y_i, called *trial vector*, by recombining the donor with the corresponding target vector. There are many kinds of crossover. The most known is the binomial crossover, but, considering previous results showed in [1], we decided to use the *interm* operator.

The *interm* crossover operator is a randomized version of the arithmetic crossover. If x_i is the target and $\bar{y}_i$ is the donor, then the trial y_i is obtained in the following way: for each component $x_i^{(h)}$ of x_i and $\bar{y}_i^{(h)}$ of $\bar{y}_i$, let $a_i^{(h)}$ be a vector of $d^{(h)}$ numbers randomly generated with a uniform distribution $[0, 1]$, then

$$y_{ij}^{(h)} = a_{ij}^{(h)} x_i^{(h)} + (1 - a_{ij}^{(h)})\bar{y}_{ij}^{(h)}$$

for $j = 1, \ldots, d^{(h)}$.

Finally, the commonly used selection operator is the one-to-one operator, that compares each trial vector y_i with the corresponding target vector x_i (by means of the fitness function) and keeps the better of them in the population of the next generation.

Another possible selection operator is the crowding operator, which was introduced in [6] and then, among others, revived in [13]. The goal in crowding is to preserve genetic diversity among population individuals, resulting in a better coverage of the search space. In this selection operator each trial vector y_i is compared to the vector $x_{clos(i)}$, that is the vector in the current population which minimizes the distance with y_i. In this way, it is unlikely that the population has similar elements and a greater level of diversity is maintained.

2.2 Cooperative Coevolution

Cooperative Coevolution (CC) has been proposed in [11] to introduce a notion of modularity in the evolutionary algorithm theory. The underlying idea was to introduce the interaction among adaptive subcomponents in order to facilitate the evolution of a complex structure.

CC is a general framework for applying Evolutionary Algorithms to large and complex problems using a divide-and-conquer strategy. The objective system is decomposed into smaller modules and each of them is assigned to a species (i.e. subpopulation). The species are evolved mostly separately with the only cooperation happening during fitness evaluation.

The use of multiple interacting subpopulations has also been explored as coevolution of species but, in this case the approach consider subpopulations evolving competing (rather than cooperating) solutions.

According to [11], in a cooperative coevolutionary algorithm each subpopulation represents a subcomponent of a potential solution, complete solutions are obtained by assembling representative members of each of the subpopulations and credit assignment at the subpopulation level is defined in terms of the fitness of the complete solutions in which the species members participate. In order to make the most of applying CC, a large problem should be decomposed into subcomponents having minimal interdependencies among different subcomponents.

Yang et al. [15] showed very good performances of their DECC-G in tackling large optimisation problems with dimensions up to 1000.

3   Algorithm

The algorithm CoDE here proposed is an extension of DENN [3] with two important additional components: a cooperative coevolutionary mechanism [9], and a crowding method [13] that penalizes too similar candidate solutions. The general algorithm is called CoDE and it is shown in Algorithm 1.

3.1   Problem Decomposition

At first, the problem has to be decomposed into smaller subcomponents that can be separately handled and evolved. We have defined two different level of decomposition: layer decomposition (CoDE-Layer) and matrix decomposition (CoDE-Matrix).

Let $\Gamma_1, \ldots, \Gamma_P$ be the P subcomponents describing the neural network we want to train, we will denote by γ_{ip} the i-th element of Γ_p and by n_p the cardinality of Γ_p.

In *decomposition by layer*, the neural network is decomposed in L layers and there is a subpopulation for each layer. The subpopulation element γ_{il} is the vector containing all the parameters in the lth layer.

In *decomposition by matrix*, each network layer is further decomposed in two matrices, one for the connection weights and one for the biases, and there is a subpopulation for each matrix. Thus, there are $2L$ subpopulations: the subpopulation elements $\gamma_{i,2l-1}$ and $\gamma_{i,2l}$ are vectors containing, respectively, the connection weights and biases in the lth layer.

3.2   Subcomponent Optimization

In this step, each subcomponent is separately evolved using DENN [1,3].

The training set TS is split in nb batches $B_0, \ldots, B_{nb-1}$ with the same size and during the evolution the networks are evaluated using only the examples belonging to one batch. The evolutionary algorithm is executed for G generations, starting from the first batch B_0. Every g_b generations, the next batch in the sequence is selected, and all the individuals of all sub-populations Γ_i are re-evaluated on that batch.

The evolution step is performed by using the mutation operator *current_to_pbest* and crossover operator *interm* [7] to create a new offspring γ'_{ip} for each element γ_{ip}. We have chosen these operators since they demonstrated the be a good combination in neural network weights optimization [2,3].

3.3   Subcomponents Recombination and Evaluation

In the evaluation of the offspring γ'_{ip}, a complete network Θ'_{ip} is built by combining γ'_{ip} with the best element β_r of each other subpopulation Γ_r, with $r \neq p$.

We denote the components of Θ'_{ip} as $[\beta_1, \ldots, \beta_{p-1}, \gamma'_{ip}, \beta_{p+1}, \ldots, \beta_P]$. Θ'_{ip} is compared, by means of the *loss function*, to $\Theta_{ip} = [\beta_1, \ldots, \beta_{p-1}, \gamma_{ip}, \beta_{p+1}, \ldots, \beta_P]$ and the better of them is selected for the next generation.

At the end of main loop, the current best solution $\Theta^* = [\beta_1, \ldots, \beta_P]$ is compared with the best solution found so far Θ^{best}.

Finally, CoDE returns the best network Θ^{best} found in all the generations.

Algorithm 1. CoDE : Cooperative Coevolutionary DENN

1: **procedure** CoDE(P, n_p, G, g_b, nb, TS)
2:    P is the number of subpopulations, n_p are the subpopulation sizes
3:    G is the number of generations; g_b is the number of generations for each batch
4:    nb is the number of batches; TS is the training set
5:    Initialize the population
6:    Extract the nb batches $B_0, \ldots, B_{nb-1}$
7:    **for** $g \leftarrow 1$ **to** G/g_b **do**
8:       Set the current batch as $B_{g \pmod{n)b}}$
9:       Evaluate all individuals of all populations
10:      **for** $z \leftarrow 1$ **to** g_b **do**
11:         **for** $p \leftarrow 1$ **to** P **do**
12:            **for** $i \leftarrow 1$ **to** n_p **do**
13:               $\gamma'_{i,p} \leftarrow$ apply mutation and crossover to $\gamma_{p,i}$
14:            **end for**
15:         **end for**
16:         **for** $p \leftarrow 1$ **to** P **do**
17:            **for** $i \leftarrow 1$ **to** n_p **do**
18:               $\Theta'_{ip} \leftarrow [\beta_1, \ldots, \beta_{p-1}, \gamma'_{ip}, \beta_{p+1}, \ldots, \beta_P]$
19:               $\Theta_{ip} \leftarrow [\beta_1, \ldots, \beta_{p-1}, \gamma_{ip}, \beta_{p+1}, \ldots, \beta_P]$
20:               **if** $f(\Theta'_{ip}) < f(\Theta_{ip})$ **then**
21:                  $\gamma_{i,p} \leftarrow \gamma'_{i,p}$
22:               **end if**
23:            **end for**
24:         **end for**
25:      **end for**
26:      Update $\beta_1, \ldots, \beta_P$
27:      Let $\Theta^* \leftarrow [\beta_1, \ldots, \beta_P]$
28:      **if** $f(\Theta^*) < f(\Theta^{best})$ **then**
29:         $\Theta^{best} \leftarrow \Theta^*$
30:      **end if**
31:   **end for**
32:   **return** Θ^{best}
33: **end procedure**

---

4 Experiments

The main purpose of the experiments is to investigate whether the presence and the granularity of the coevolution affect the performances of the algorithm. Another important aspect to be considered is the effect of employing the batch system. Finally, we study if the crowding selection works, and if it is the right choice to improve the quality of the solution.

In order to make fair comparisons, we also implemented a third version of the algorithm CoDE, called CoDE-Net, which has just one subpopulation representing the whole network and does not use coevolution.

The experiments were performed on a machine equipped with 16 GB of RAM and AMD 2700X CPU.

4.1 Datasets

The datasets used in our work are the same used in [14]: the *Wisconsin breast cancer* (WBC) dataset, the *Epileptic Seizure Recognition* (ESR) dataset and the *Human Activity Recognition* (HAR) dataset. The characteristics of each dataset are shown in Table 1 following an increasing complexity order. More precisely, WBC and ERS are two binary classification problems, with the latter being much larger than the former both in terms of number of features and samples; HAR is a multiclass classification problem with a even higher size, hence it is the most complex problem among these three datasets.

Table 1. Datasets used in the experiments.

Dataset	Class	Features	Instances	Parameters
WBC	2	30	569	1652
ESR	2	178	4600	9052
HAR	6	561	7144	28406

Moreover, in order to show how the complexity increases, the last column of the Table contains Table 1 the overall number of the network parameters for each dataset.

4.2 Parameters

CoDE has 4 parameters to be set: the subpopulation size n_p, the number of generations G, the number of generations for each batch g_b, and the number of batches nb. The number of subpopulations P is determined by the network structure.

n_p is set to 20; the network is trained for 5000 generations on WBC, 1500 on ESR, and 2500 on HAR; nb is set to 100 on WBC, 500 on ESR and HAR; finally, g_b is set to 3 on the first dataset, to 5 on the second one and 10 on the last one (as in [14]).

The network layout is the same for all the experiments, i.e. a MLP with one hidden layer composed of 50 neurons with the sigmoid activation function.

4.3   Experimental Results

In this section, some experimental results are presented and discussed.

In Table 2 the results obtained by all the CoDE settings without crowding on all the datasets are reported, where the average accuracy obtained over 10 runs and the corresponding standard deviations are shown. From these we can conclude that the best setting for the datasets WBC, ESR and HAR is CoDE-Matrix, i.e. CoDE with the highest level of granularity.

Table 2. Results without crowding

Algorithm	WBC	ESR	HAR
CoDE-Net	90.23 ± 2.47	80.91 ± 3.58	35.05 ± 6.62
CoDE-Layer	96.86 ± 1.11	86.37 ± 0.20	48.00 ± 3.59
CoDE-Matrix	94.70 ± 4.22	**88.66 ± 0.91**	67.17 ± 4.85
CoDE-Net ($nb = 1$)	93.64 ± 1.70	83.10 ± 4.23	50.39 ± 7.05
CoDE-Layer ($nb = 1$)	94.11 ± 0.96	86.52 ± 0.89	65.36 ± 1.06
CoDE-Matrix ($nb = 1$)	**97.88 ± 1.37**	87.59 ± 1.00	**71.91 ± 1.45**

It is worth noticing that, differently from [1], the use of the batching system does not give positive contribution in terms of accuracy. It is probably due to the dimensions of the tested datasets, that are not so big to require a batching system.

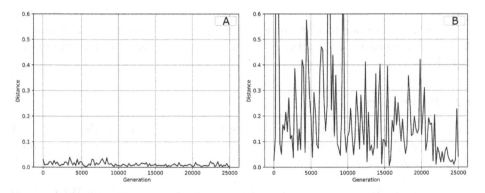

Fig. 1. Average values of the distances among the population elements for each generation in the case of CoDe_Layer (BS) on the HAR dataset for layer 0 without crowding (plot A) and with crowding (plot B)

During this test phase we noted that after a certain amount of generations, the evolutionary process often gets stuck because the heterogeneity of the population suddenly falls down, i.e. all the population members tend to float around a particular individual, likely a local minimum.

So, we decided to implement the crowding method described in Sect. 2.1 and we studied how the population is affected by it.

In Fig. 1 an example of the average distances between the elements of the whole population for each generation on the HAR dataset are shown. The distances when CoDE is configured without and with crowding are shown, respectively, in charts A and B. As visible in Fig. 1, crowding significantly increases the average distance between the elements of the population, guaranteeing the population diversity.

Table 3. Results with crowding

Algorithm	WBC	ESR	HAR
CoDE-Net	88.11 ± 0.91	78.80 ± 4.51	31.61 ± 7.28
CoDE-Layer	96.07 ± 1.99	88.06 ± 1.03	46.98 ± 8.22
CoDE-Matrix	94.82 ± 2.53	88.18 ± 0.66	$\mathbf{68.03 \pm 4.96}$
CoDE-Net ($nb = 1$)	91.76 ± 2.35	84.43 ± 0.80	58.29 ± 11.08
CoDE-Layer ($nb = 1$)	93.33 ± 0.55	86.81 ± 1.03	65.45 ± 11.66
CoDE-Matrix ($nb = 1$)	$\mathbf{97.17 \pm 1.20}$	$\mathbf{88.52 \pm 1.39}$	66.83 ± 7.05

In order to understand the contribution of crowding in terms of classification accuracy, we run another set of test across all the three datasets. In Table 3, the results are shown and we can note that, also in this case, CoDE-Matrix turns out to be the best configuration. Moreover, we can also note that, despite the use of crowding seems to be of help to maintain the population diversity, the training process does not seem to be in general positively affected.

5   Conclusions

In this work, we introduced an algorithm based on Cooperative Coevolutionary Differential Evolution (CoDE) to train neural networks, and we studied how this algorithm works on different classification datasets.

The experiments suggest that the best accuracy levels are obtained with the CoDE-matrix configuration, i.e. with the finest level of granularity for problem decomposition among the tested ones.

Moreover, we studied also the contribution of a batching system and a crowding method with respect to the base algorithm proposed.

In this case we can note that, despite of lower computational time, the batch system does not help in term of accuracy. This is not surprisingly considering the dimensions of datasets. Its impact on larger problems still has to be investigated. Finally, about the use of crowding, we can say that it is very effective to maintain population diversity even if this does not correspond to a significant improvement of the classification accuracy. Moreover crowding selection increases convergence times, suggesting that a larger number of generations is needed to find better solutions; further tests are hence required to understand whether crowding is a good choice or not to overcome the problem of population stagnation.

As a future lines of research, we are planning to test other methods to reduce the issue of population stagnation and to extend CoDE with higher levels of granularity for problem decomposition, i.e. create cooperative subpopulations by groups of neurons or single neurons.

References

1. Baioletti, M., Di Bari, G., Milani, A., Poggioni, V.: Differential evolution for neural networks optimization (2019, To appear)
2. Baioletti, M., Belli, V., Di Bari, G., Poggioni, V.: Neural random access machines optimized by differential evolution. In: Ghidini, C., Magnini, B., Passerini, A., Traverso, P. (eds.) AI*IA 2018. LNCS (LNAI), vol. 11298, pp. 307–319. Springer, Cham (2018). https://doi.org/10.1007/978-3-030-03840-3_23
3. Baioletti, M., Di Bari, G., Poggioni, V., Tracolli, M.: Can differential evolution be an efficient engine to optimize neural networks? In: Nicosia, G., Pardalos, P., Giuffrida, G., Umeton, R. (eds.) MOD 2017. LNCS, vol. 10710, pp. 401–413. Springer, Cham (2018). https://doi.org/10.1007/978-3-319-72926-8_33
4. Das, S., Mullick, S.S., Suganthan, P.: Recent advances in differential evolution - an updated survey. Swarm Evol. Comput. **27**, 1–30 (2016)
5. Das, S., Suganthan, P.N.: Differential evolution: a survey of the state-of-the-art. IEEE Trans. Evol. Comput. **15**(1), 4–31 (2011)
6. De Jong, K.A.: Analysis of the behavior of a class of genetic adaptive systems (1975)
7. Eltaeib, T., Mahmood, A.: Differential evolution: a survey and analysis. Appl. Sci. **8**(10) (2018). https://www.mdpi.com/2076-3417/8/10/1945
8. Morse, G., Stanley, K.O.: Simple evolutionary optimization can rival stochastic gradient descent in neural networks. In: GECCO 2016, pp. 477–484. ACM (2016)
9. Olorunda, O., Engelbrecht, A.P.: Differential evolution in high-dimensional search spaces. In: Proceedings of CEC, 2007, pp. 1934–1941 (2007)
10. Piotrowski, A.P.: Differential evolution algorithms applied to neural network training suffer from stagnation. Appl. Soft Comput. **21**, 382–406 (2014)
11. Potter, M.A., De Jong, K.A.: A cooperative coevolutionary approach to function optimization. In: Davidor, Y., Schwefel, H.-P., Männer, R. (eds.) PPSN 1994. LNCS, vol. 866, pp. 249–257. Springer, Heidelberg (1994). https://doi.org/10.1007/3-540-58484-6_269
12. Prellberg, J., Kramer, O.: Limited evaluation evolutionary optimization of large neural networks. In: Trollmann, F., Turhan, A.-Y. (eds.) KI 2018. LNCS (LNAI), vol. 11117, pp. 270–283. Springer, Cham (2018). https://doi.org/10.1007/978-3-030-00111-7_23
13. Thomsen, R.: Multimodal optimization using crowding-based differential evolution. In: Proceedings of CEC 2004, vol. 2, pp. 1382–1389 (2004)
14. Yaman, A., Mocanu, D.C., Iacca, G., Fletcher, G., Pechenizkiy, M.: Limited evaluation cooperative co-evolutionary differential evolution for large-scale neuroevolution. In: Proceedings of GECCO 2018, pp. 569–576. ACM (2018)

15. Yang, Z., Tang, K., Yao, X.: Large scale evolutionary optimization using cooperative coevolution. Inf. Sci. **178**(15), 2985–2999 (2008)
16. Zhang, X., Clune, J., Stanley, K.O.: On the relationship between the OpenAI evolution strategy and stochastic gradient descent. arXiv preprint arXiv:1712.06564 (2017)

Design and Evaluation of a Heuristic Optimization Tool Based on Evolutionary Grammars Using PSoCs

Bernardo Vallejo Mancero[1,2]([⊠]), Mireya Zapata[3][iD], Liliana Topón - Visarrea[3], and Pedro Malagón[1]

[1] ETSI Telecomunicación, Universidad Politécnica de Madrid,
Av. Complutense 30, 28040 Madrid, Spain
b.vallejo@alumnos.upm.es, malagon@die.upm.es
[2] Instituto Superior Tecnológico Central Técnico,
Av. Isaac Albéniz E4-15 y El Morlán, Quito, Ecuador
[3] Research Center in Mechatronics and Interactive Systems,
Universidad Indoamérica, Machala y Sabanilla, Quito, Ecuador
{mireyazapata,blancatopon}@uti.edu.ec

Abstract. Currently, the evolutionary computing techniques are increasingly used in different fields, such as optimization, machine learning, and others. The starting point of the investigation is a set of optimization tools based on these techniques and one of them is called evolutionary grammar [1]. It is a evolutionary technique derived from genetic algorithms and used to generate programs automatically in any type of language.

The present work is focused on the design and evaluation of hardware acceleration technique through PSoC, for the execution of evolutionary grammar. For this, a ZYNQ development platform is used, in which the logical part is used to implement factory modules and independents hardware blocks made up of a soft-processor, memory BRAM, and a CORDIC module developed to perform arithmetic operations. The processing part is used for the execution of the algorithm. Throughout the development, the procedures and techniques used for hardware and software design are specified, and the viability of the implementation is analyzed considering the comparison of the algorithm execution times in Java versus the execution times in Hardware.

Keywords: Evolutionary grammar · FPGA · PSoC · CORDIC

1 Introduction

Optimization problems are found in several areas, such as mathematics, computer science, engineering, and even everyday life [2]. It is so, whenever is about willing to improve profits, energy consumption, or minimizing errors and use of resources, among others we talk about optimization problems.

© Springer Nature Switzerland AG 2020
F. Cicirelli et al. (Eds.): WIVACE 2019, CCIS 1200, pp. 100–112, 2020.
https://doi.org/10.1007/978-3-030-45016-8_11

Today the use of techniques based on exact resolution algorithms, which end in a finite number of steps, has been relegated to heuristic techniques, which are non-traditional methods, which through the generation of candidates for possible solutions deal with solving problems on different fields. Genetic algorithms are part of heuristic techniques, these are algorithms that emulate natural evolution and their goal is the searching for optimization of solutions to different problems, solutions that have better qualities than other ones existing [2].

Evolutionary grammar (EG) is a technique derived from genetic algorithms [3]. This technique is based on the concept of genetic recombination, and can be used to automatically generate programs in any language according to the specified grammar [12]. The GE uses "Backus-Naur" notation (BNF), by means of which a context-free grammar is defined where a series of rules production compound of terminal and non-terminal elements is established [4,11].

It is important to highlight that software solutions that implement heuristic techniques present an inherently sequential processing, with which execution times are directly proportional to the population and the complexity of the algorithm process. It can cause a high demand for processor resources along with long execution times, limiting the scalability of the application [5].

The objective of this work is to design and evaluate hardware acceleration options in an optimization tool based on evolutionary grammars developed in Java [6], making modifications in different classes and replacing them with hardware blocks, to determine the benefits of including acceleration using a platform PSoC.

The execution of the Java GE optimization tool is divided into 5 stages: problem definition, operator definition, algorithm initialization, evaluation and execution.

- Definition of the problem: The declaration of the number of individuals and generations to be evaluated is made. Here the location of the files that contain the description of the grammar, the training data set, and the compiler configuration is defined.
- Definition of operators: According to the optimization problem, the type of operators to be used is specified, which can be mutation, crossing, comparison and selection.
- Initialization: A starting population is taken as starting point, created with a random set of solutions, where each individual is formed by a genotype, which is a sequence of values (genes) that constitute the coded information of the individual. A phenotype is obtained from a genotype, which is the expression or program that is created based on a defined grammar, each individual becomes a tree, which once decoded, can contain arithmetic expressions, programs, logical formulas, and others.
- Evaluation: Each individual of the population is evaluated independently with respect to an adaptation function (fitness), which is defined by a set of inputs and outputs training data.
- Execution: The individuals with the best results can pass intact to the next generation or apply techniques of genetic operations (mutation and crossing), until obtaining a new population, which will be evaluated again.

The described process is repeated iteratively until the optimal solution is obtained, or until the limit of established generations is reached.

In the evaluation stage, each phenotype must be evaluated for each of the training values. The number of operations is equal to the product of the number of individuals by the number of training values, which in large populations represents a significant time because in the original solution it is carried out sequentially (individual after individual). The objective of the present design is to parallelize this process through the development of independent hardware blocks, which represent each of the individuals. This instances are reprogrammed each generation according to the new solutions generated.

1.1 PSoC Platform

It is a platform based on the Zynq architecture of Xilinx, which integrates an FPGA and a dual-core ARM Cortex-A9 processor, allowing a Hardware Software co-design offering the flexibility of the hardware and the simplicity of the software. This architecture is made up of two parts: Processing System (PS) implemented with an ARM processor and Programmable logic (PL) based on FPGA. The communication between PS and PL is done through AXI interfaces (advanced extensible interface). The AXI specifications describe a master-slave communication interface that allows the exchange of information between IP modules [7].

The following section describes the design carried out, which is divided into two parts, one dedicated to the Hardware design and the other to the Software design, Sect. 3 shows the implementation of the design and the resources used. A description of the results obtained is presented in Sect. 4, the results are discovered in Sect. 5, finally, the conclusions and future work are presented in Sect. 6.

2 Design

For the design of the solution proposed in this work, it is necessary to modify the GE in Java to include the hardware acceleration in the execution sequence of the algorithm. The modifications is carried out in the initialization stage in the generation of the phenotype, where a change is made to the grammar in Java so that the result can be compiled from the C language in order to translate instructions to hexadecimal code. The new compiler must be executed from the original library and be able to generate the algorithm instructions (hexadecimal code) to be evaluated that can be interpreted by the PSoC. The evaluation stage is implemented in both PS and PL domains as illustrated in Fig. 1 and whose functions are detailed below.

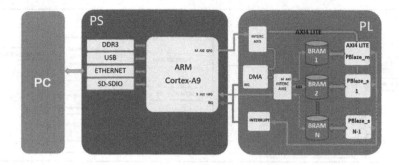

Fig. 1. Hardware diagram and connections

2.1 Design PL

It is made up of independent hardware blocks that process the instructions generated in Java and determine the fitness function of each generation. Each block called PBlaze consists of a memory, the KCPSM6 [8] processor of the Picoblaze microcontroller and communication buses. Other hardware blocks used are the DMA, BRAM blocks, interruption block and interconnection blocks.

The description of the design is made according to the order of the flow of information for each of the modules.

Direct Access to Memory. The generated instructions are written in the memory of each PBlaze. To reduce the use of the CPU, a direct memory access block (DMA) [9] is implemented, the chosen IP block is a central memory access AXI controller (AXI-CDMA), which allows the transfer of data to different memories mapped within an address list. The CDMA configuration is done through an AXI4-lite slave bus connected to the ARM processor.

Memory. The selected memories are RAM blocks (BRAM)of 8KB capacity, which allows a maximum of 2048 instructions of up to 32 bits for each PBlaze. They are configured in "True Dual-port RAM" mode. Each of its ports operate independently with reading and writing capability. Port A is connected to the CDMA and port B to the PBlaze module.

PBlaze Evaluation Module. Continuing with the process, when the writing of the instructions is concluded, the evaluation stage is passed. The same data must be sent from the PS and received by each of the PBlaze. Two types of PBlaze modules are developed, one master (PBlaze_m) and another slave (PBlaze_s). These are shown in Fig. 2 (a) and (b) respectively. The two modules are constituted by a data sending bus, a memory access bus, a KCPSM6 processing unit and a CORDIC module for arithmetic operations. The two types differ because the master PBlaze module uses an AXI communication bus and the slave PBlaze module uses a simple data reception bus.

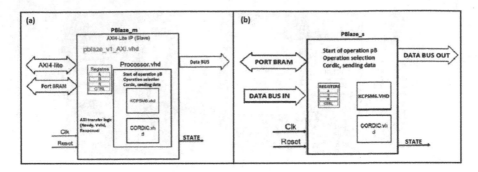

Fig. 2. IP core architecture custom PBlaze master (a) and PBlaze slave (b)

The connection between modules has a cascade structure, where in each clock cycle the data is forwarded from one of the PBlaze to the next, through the data transmission bus, the process is repeated until all the modules receive the data. With the design it is not necessary that each module is connected to the PS block, which represents time and area savings.

To perform the GE evaluation, the data received from PS is sent to the processing unit in order to generate a response, which is compared with the original response. The error is determined by adding the amount of values that do not agree with the expected result. This error is finally the fitness of the solution that is recorded in the same BRAM memory block waiting to be read by the PS.

For its operation, 4 registers are implemented, each one with 32 bits: the first three registers process decimal values with sign in fixed point, the accepted range of values is:

$$- 524288 \leq n \geq 524287.9999 \tag{1}$$

The last, serves as configuration record and Start-up.

BUS AXI4-Lite. It is the communication bus selected to be implemented in the evaluation module [7]. It allows the transfer of a data of 32 bits for each transaction, with 2 channels for reading and 3 for writing. Each channel has its own "ready" and "valid" signals, with functions that depend on the source and destination of the message.

KCPSM6. It fulfills the function of generating a response based on the interpretation of the instructions stored in each BRAM block, and of the training input data. The ports in the KCPSM6 processor are only 8 bits, but thanks to "port_id" we can work with up to 256 I/O ports. Using this feature, 32-bit data is processed, mapping ports and registers.

Each record is divided into bytes and each byte is associated with a port_id, so, to read a complete data, it is necessary to read 4 instructions from the input port changing the port_id. The same criteria is applied to the output data, which is connected to two 32-bit output registers, used to perform arithmetic operations using a CORDIC module.

CORDIC Module. It is a module developed to perform arithmetic operations not included in the KCPSM6; the operations to be carried out, as well as the loading of the data records, is the responsibility of the KCPSM6 module; the operations implemented are multiplication and division, Sine, cosine, tangent, logarithm base 10, exponential, and power.

In the case of multiplication, which is one of the most used operations, a solution with a higher response speed is used, for which the multiplication module of Xilinx "Multiplier" is implemented.

For the other operations, the CORDIC algorithm [10] is used to adjust it according to the application's need and data size. The algorithm performs a vector rotation as a sequence of smaller successive rotations, for the implementation of the algorithm three variables are necessary (x, y, z) and the generalized formula proposed by John Stephen Walther of Hewlett-Packard was used:

$$x_i + 1 = x_i + m\alpha_i y_i 2^{-i} \tag{2}$$

$$y_i + 1 = y_i - \alpha_i x_i 2^{-i} \tag{3}$$

$$z_i + 1 = z_i + \alpha_i f(2^{-i}) \tag{4}$$

where:
   m = 1 circular rotation, 0 linear rotation, and −1 hyperbolic rotation.
   α = Direction of rotation (1 or −1).
   $f(2^{-i}) = arctg(2^{-i})$ circular rot., 2^{-i} linear rot., $arctgh(2^{-i})$ hyperbolic rot.

Interruption Block. It is another customized IP module in which all the status signals of the PBlazes modules are connected. Its function is to notify the PS that all PBlazes blocks have finished processing the data sent so that a new data can be sent. The signal is connected to one of the PS interrupt ports.

2.2   Design PS

C language is used in a bare metal application; its main functions are:

- Addressing and sending the instructions of each PBlaze module to the CDMA block.
- Sending training data to the PBlaze modules through the AXI bus.
- Fitness reading corresponding to each evaluated PB.
- Recognition of interruptions.

The blocks connected to the PS have physical addresses, included at the beginning of the application, that allow the configuration of the block and the data transfer to the PL and vice versa. The PL blocks which are mapped in memory are: the IP CDMA and the IP PBlaze_m, blocks used for the transferring instructions and training data respectively. Other necessary addresses are those that the CDMA uses to recognize the memories BRAM and DDR. Finally, other parameters used are the number of training data and the port on which the interruption block is connected.

The instructions for each individual are previously generated and stored in a ROM.c file. Which is transferred directly to the DDR memory of the PS. The CDMA points to the address of the DDR memory as a source and transfers the amount of bytes set to the destination memory which can be any of the BRAMs of the PBlaze modules.

Recorded all the instructions begin to transfer the data through the AXI bus to the PBlaze_m block. One by one the data is sent, including the control signal. Each time that the processing of a data is finished, an interruption is generated that enables the sending of a new data, the process is repeated until all the data has been transferred. Once the data has been sent, the CDMA reads the fitness stored in the first position of the BRAM blocks of each PBlaze and transfers it to the DDR memory for use in the following stages of the evolutionary algorithm.

3 Implementation

The proposed system consists of 25 PBlaze modules (1 pblaze_m and 24 pblaze_s). One stage of the implemented design is observed in Fig. 3. The implementation is done on a Xilinx Zynq7000 SoC ZC706 Evaluation Kit. Figure 4 is a summary of the resources used, obtained from the implementation stage in Vivado. It is observed that the critical data in the implementation is the quantity of Slice LUTs used, for a population of 25 individuals it reaches 28%, the critical parameter determines the maximum size of the generation that can be implemented. Carried out a deeper examination determines the amount of resources that each individual occupies. When evaluating the most used resource, it is obtained that each individual occupies 0.82% of the maximum capacity of the FPGA according to the data shown in Table 1. Considering the remaining capacity, up to 100 individuals could be implemented per generation.

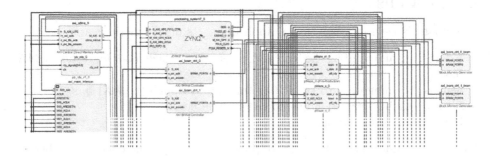

Fig. 3. Hardware diagram of the developed solution

The most used resources are the Look Up Table (LUT), mainly due to the logic used for the implementation of the Cordic algorithm. To optimize resources only the operations necessary for the use of the grammar are considered.

Resource	Utilization	Available	Utilization %
LUT	61338	218600	28.06
LUTRAM	2470	70400	3.51
FF	44785	437200	10.24
BRAM	56	545	10.28
DSP	100	900	11.11
BUFG	2	32	6.25

Fig. 4. Total resources used

Table 1. Resources used by each individual.

Module	Total LUT	Logic LUT	LUT RAM	FFs	RAM B36	DSP 48
pblaze_m_0	1609(0.74)	1585(0.72)	24(0.03)	934(0.21)	0(0.0)	4(0.4)
(pBlaze_S_AXI_insz)	53(0.02)	53(0.02)	0(0.00)	206(0.05)	0(0.0)	0(0.0)
(Proccesor_v1)	1556(0.72)	1532(0.70)	0(0.00)	728(0.16)	0(0.0)	4(0.4)
axi_bram_ctrl_0	176(0.08)	176(0.08)	0(0.00)	183(0.08)	0(0.0)	0(0.0)
axi_bram_ctrlbram	0(0.00)	0(0.00)	0(0.00)	0(0.00)	2(0.4)	0(0.0)
TOTAL	1785(0.82)	1761(0.80)	24(0.03)	1117(0.3)	2(0.4)	4(0.4)
pblaze_s_0	1520(0.70)	1496(0.68)	24(0.03)	796(0.18)	0(0.0)	4(0.4)
axi_bram_ctrl_0	176(0.08)	176(0.08)	0(0.00)	183(0.08)	0(0.0)	0(0.0)
axi_bram_ctrlbram	0(0.00)	0(0.00)	0(0.00)	0(0.00)	2(0.4)	0(0.0)
TOTAL	1696(0.78)	1672(0.76)	24(0.03)	979(0.26)	2(0.4)	4(0.4)

For the implementation of the PS design, the Vivado SDK is used, where the main.c, ROM.h and data.h files are defined. In the first one is the declaration of the application. The instructions to be recorded in each BRAM memory are stored in ROM.h and finally the training data converted to fixed point are stored in data.h.

4   Results

The evaluation of 25 solutions (phenotypes) generated by the application in Java is performed and the times obtained in the PC and the FPGA are compared. First, it is necessary to convert each phenotype in its equivalent in assembly language and later to hexadecimal code to record each one of the memories.

The tests are carried out on the Zynq 7000 development card and on a PC with the characteristics of Table 2.

Table 2. Description of PC and PSoC features.

Device	Features	Frequency
PC	Processor: Intel(R) Core(TM) i7-8750H (12 CPUs) 8th Gen	2.20 GHz
Zynq-7000	Processor: ARM Cortex-A9 (2 cores)	650 MHz
	Programmable logic	125 MHz

The phenotypes evaluated comply with the following structure:

``double result = 0.0; if (``<LogExpr>'') {result = 0.0;}
else {result = 1.0;} return result;''

Corresponding to the grammar used. A total of 300 data is used in the evaluation, each data is formed by two input values and one result. At the end, the fitness and the total time that takes the evaluation of each generation is obtained.

The evaluation of 10 generations of 25 phenotypes in the PC presents the results shown in Fig. 5a, it should be noted that the time measured corresponds only to the evaluation stage, the best fitness is 18, the minimum time is 2153.28 μs and the maximum is 3142.96 μs. The total time used to evaluate all possible solutions is equal to the sum of the individual times of each generation.

For the test on the Zynq-7000 card, the generation that presented the worst evaluation time is chosen, instructions are passed in hexadecimal code and the bare metal application is run. The results obtained are shown in Fig. 5b.

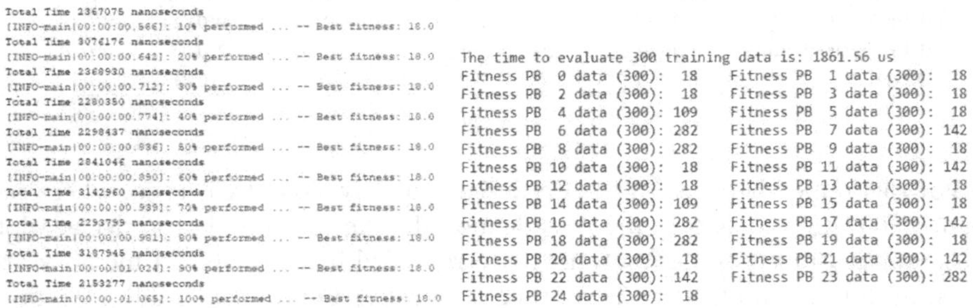

(a) Software runtime

(b) Hardware runtime

Fig. 5. Execution of the evaluation stage of GE

The best fitness is 18 and the total time to evaluate the generation is 1861.56 μs, unlike the execution in the PC, in which the executions are sequential. In the FPGA, the time it takes each generation is determined by the maximum time it takes for one of the phenotypes to be evaluated.

A summary of the data obtained and a projection can be observed in Table 3, in which the time that takes 10 generations to be evaluated in the PC is added and the solution in the Zynq card is approximated to a greater number of generations.

Table 3. Comparison of software and hardware execution.

Generation number	Software solution JAVA	Hardware solution Zynq-7000	(%)
1	3187.95 μs	1861.56 μs	41.55%
10	26.01 ms	18.61 ms	28.37%

In the case of 25 solutions, when comparing the results, it is obtained that the times of the application running in the FPGA with respect to the solution in the PC are smaller. By increasing the number of individuals per generation these times tends to continue to improve.

5 Discussion

Currently, evolutionary grammars are being applied in different projects, such as: automatic generation of neural networks [14], business analysis [15], short-term blood glucose prediction model [16], among others. The implementation of evolutionary techniques has traditionally been done in different types of language [13], when these are executed on a CPU, the speed is one of the factors that developers consider to decide on one type or another. Based on a Benchmarking for evolutionary languages, it was demonstrated that Java is the language with the best results [17].

In past years, the use of FPGAs in high-performance computers as replacement of processors was investigated [18]. Studies showed the feasibility, however, at that time the cost factor was a limitation that, at present, with the advance and improvement of technology, has been reduced.

The use of an FPGA in the evaluation stage of the algorithm seeks to further improve the execution speed, through the parallelism offered by these devices. Parallelism that has been shown to give better results in tests carried out with genetic algorithms, for example, in the resolution of the TSP (Traveling Salesman Problem), a problem that seeks to find the best route between different nodes. It was determined that although for this type of problem with multiple accesses to memory, the use of the FPGA was not the most appropriate, for other types of genetic algorithms where there is a greater amount of computation and less access to memory for the fitness function is a good alternative [19], characteristics that the current evolutionary grammar presents when processing different instructions for each individual and facilitate obtaining the fitness function through the sum of values that do not agree with the expected result.

Comparing the execution of Software versus Hardware, taking into account only the evaluation stage and only 10 generations, it is obtained that overall for a population of 25 individuals the times are reduced by an average of 28%. The results can be improved if more individuals are implemented in the FPGA, or if the amount of data to be evaluated is increased.

Applying Amdahl's law [20] described by Ec (5) can obtain the total increase in speed thanks to the improvement, for the analysis it is considered 10 generations and a population of 100 individuals, maximum allowed by the FPGA, which maintain the same time as the worst case evaluated in hardware. The following values are considered:

- Fm: fraction of the time in which the improvement is made. Taking into account that the execution time of the algorithm in software is 350 ms and the time that takes the evaluation stage is 104.04 ms, Fm is equal to 29.73%
- Sm: Increase in speed during improvement. The increase is equal to the improvement in time comparing the execution in software (104.04 ms) versus hardware (18.63 ms), which is approximately 5.58 times faster.

$$S = 1/[(1 - Fm) + Fm/Sm] \tag{5}$$

$$S = 1/[(1 - 0.297) + 0.297/5.58] \tag{6}$$

$$S = 1.32 \tag{7}$$

As summary, the implementation of the evaluation stage in an FPGA with the current design can return the algorithm up to 1.32 times faster than software execution.

6   Conclusion and Future Work

The hardware design together with the standalone application demonstrated that, in the case of evaluating the same number of individuals, the designed solution had better results, in terms of execution time. Although, the use of the Picoblaze microcontroller offers advantages in terms of size and ease of integration, its great disadvantage is that it is an 8-bit processor with a limited ISA. This causes the number of instructions necessary to solve any problem to be 4 times larger than it would be with a 32-bit processor, which in turn increases the execution time. Also, the syntax represents a bottleneck, due to the need of external compilers in the case of passing from one type of language to another.

The resources offered by the PSoC platform is a factor that greatly limits the real capacity of the solution, and increases the effort and time in the search for an optimal solution that fits the platform.

As future work, the change of design in the evaluation module by a custom ALU that fits the needs of the problem, including the syntax of the phenotype itself is considered. Others, a little more ambitious are the modification of the grammar syntax to fit the hardware solution. It is also possible to develop a new own library for embedded systems.

In this work we have focused on the evaluation phase, but future efforts can be focused on the development of hardware acceleration techniques for other stages of the algorithm, such as the compilation or generation of new individuals.

Finally, we can consider the solution adopted for other algorithms and applications in the field of evolutionary techniques and others in which a high degree of parallelism is required.

References

1. Nicolau, M., Agapitos, A.: Understanding grammatical evolution: grammar design. In: Ryan, C., O'Neill, M., Collins, J.J. (eds.) Handbook of Grammatical Evolution, pp. 23–53. Springer, Cham (2018). https://doi.org/10.1007/978-3-319-78717-6_2
2. Kramer, O.: Genetic Algorithm Essentials. SCI, vol. 679. (2017). https://doi.org/10.1007/978-3-319-52156-5
3. De Silva, A.M., Leong, P.H.W.: Grammatical evolution. SpringerBriefs Appl. Sci. Technol. **5**, 25–33 (2015). https://doi.org/10.1007/978-981-287-411-5_3
4. O'Neill, M., Brabazon, A.: Grammatical swarm: the generation of programs by social programming. Nat. Comput. **5**, 443–462 (2006). https://doi.org/10.1007/s11047-006-9007-7
5. Le Goues, C., Yoo, S. (eds.): SSBSE 2014. LNCS, vol. 8636. Springer, Cham (2014). https://doi.org/10.1007/978-3-319-09940-8
6. Colmena, J.: HEuRistic optimization (2016). GitHub repositor. https://github.com/jlrisco/hero
7. Xilinx Inc: AXI reference guide UG761 (v13.1). 761 (2011)
8. Chapman, K.: PicoBlaze for Spartan-6, Virtex-6, 7-Series, Zynq and UltraScale Devices (KCPSM6). 1–24 (2014)
9. Dma, A.X.I.: Table of contents. Nippon Ronen Igakkai Zasshi. Japanese J. Geriatr. **56**, Contents1-Contents1 (2019). https://doi.org/10.3143/geriatrics.56.contents1
10. Volder, J.: The CORDIC computing technique, pp. 257-261 (2008). https://doi.org/10.1145/1457838.1457886
11. Ryan, C., O'Neill, M., Collins, J.J.: Introduction to 20 years of grammatical evolution. In: Ryan, C., O'Neill, M., Collins, J.J. (eds.) Handbook of Grammatical Evolution, pp. 1–21. Springer, Cham (2018). https://doi.org/10.1007/978-3-319-78717-6_1
12. Lourenço, N., Assunção, F., Pereira, F.B., Costa, E., Machado, P.: Structured grammatical evolution: a dynamic approach. In: Ryan, C., O'Neill, M., Collins, J.J. (eds.) Handbook of Grammatical Evolution, pp. 137–161. Springer, Cham (2018). https://doi.org/10.1007/978-3-319-78717-6_6
13. Grifoni, P., D'Ulizia, A., Ferri, F.: Computational methods and grammars in language evolution: a survey. Artif. Intell. Rev. **45**(3), 369–403 (2015). https://doi.org/10.1007/s10462-015-9449-3
14. Assuncao, F., Lourenco, N., Machado, P., Ribeiro, B.: Automatic generation of neural networks with structured Grammatical Evolution. In: Proceedings of the 2017 IEEE Congress on Evolutionary Computation CEC 2017, pp. 1557–1564 (2017). https://doi.org/10.1109/CEC.2017.7969488
15. Borlikova, G., Smith, L., Phillips, M., O'Neill, M.: Business analytics and grammatical evolution for the prediction of patient recruitment in multicentre clinical trials. In: Ryan, C., O'Neill, M., Collins, J.J. (eds.) Handbook of Grammatical Evolution, pp. 461–486. Springer, Cham (2018). https://doi.org/10.1007/978-3-319-78717-6_19
16. Contreras, I., Bertachi, A., Biagi, L., Oviedo, S., Vchí, J.: Using grammatical evolution to generate short-term blood glucose prediction models. In: CEUR Workshop Proceedings, vol. 2148, pp. 91–96 (2018)
17. Merelo, J.J., et al.: Benchmarking languages for evolutionary algorithms. In: Squillero, G., Burelli, P. (eds.) EvoApplications 2016. LNCS, vol. 9598, pp. 27–41. Springer, Cham (2016). https://doi.org/10.1007/978-3-319-31153-1_3

18. Craven, S., Athanas, P.: Examining the Vi-ability of FPGA Supercomputing. EURASIP J. Embed. Syst. **93652** (2007). https://doi.org/10.1155/2007/93652
19. Vega-Rodríguez, M.A., Gutiérrez-Gil, R., Ávila-Román, J.M., Sánchez-Pérez, J.M., Gómez-Pulido, J.A.: Genetic algorithms using parallelism and FPGAs: the TSP as case study. In: Proceedings of the International Conference on Parallel Processing Workshop 2005, pp. 573–579 (2005). https://doi.org/10.1109/ICPPW.2005.36
20. Hill, M.D., Marty, M.R.: Amdahl's law in the multicore era. Computer (Long. Beach. Calif) **41**, 33–38 (2008). https://doi.org/10.1109/MC.2008.209

How Word Choice Affects Cognitive Impairment Detection by Handwriting Analysis: A Preliminary Study

Nicole Dalia Cilia[(✉)], Claudio De Stefano, Francesco Fontanella,
and Alessandra Scotto di Freca

Department of Electrical and Information Engineering (DIEI),
University of Cassino and Southern Lazio,
Via G. Di Biasio 43, 03043 Cassino, FR, Italy
{nicoledalia.cilia,destefano,fontanella,a.scotto}@unicas.it

Abstract. In this paper we present the results of a preliminary study in which we considered two copy tasks of regular words and non-words, collecting the handwriting data produced by 99 subjects by using a graphic tablet. The rationale of our approach is to analyze kinematic and pressure properties of handwriting by extracting some standard features proposed in the literature for testing the discriminative power of non-words task to distinguish patients from healthy controls. To this aim, we considered two classification methods, namely Random Forest and Decision Tree, and a standard statistical ANOVA analysis. The obtained results are very encouraging and seem to confirm the hypothesis that machine learning-based analysis of handwriting on the difference of Word/Non-Word tasks can be profitably used to support the Cognitive Impairment diagnosis.

Keywords: Word and Non-Word · Handwriting · Classification algorithm · Cognitive Impairment

1 Introduction

Traditional theories have postulated that cognitive processing and the motor system were functionally independent so that one movement was the end result of cognitive processing. However, it is increasingly evident that the two systems and their relations are much more complex than previously imagined. The existence of such motor traces of the mind represents a great leap forward in research; researchers have often been forced to look at the "black box" of cognition through indirect, off-line observations, such as reaction times or errors. A serious drawback of this approach is that behavioral results provide little information on how the cognitive process evolves over time and how multiple processes can converge and guide final responses. The motor activities have been

This work is supported by the Italian Ministry of Education, University and Research (MIUR) within the PRIN2015-HAND project.

investigated with increasing frequency in recent decades. In this context, the knowledge within cognitive sciences has specialized in many application fields. One of the most prolific is certainly the medical one in which both predictive and rehabilitative research converge. For example, hand movements offer continuous flows of output that can reveal the dynamics being developed and, according to a growing body of literature, can be argued that hand movements can provide, with a high degree of fidelity, traces of the mind.

Furthermore, recently, researchers have shown that some motor activities can represent good indices of neurodegenerative disease. For example, patients affected by Cognitive Impairment (CI) exhibit alterations in the spatial organization and poor control of fine movements. This implies that, at least in principle, some diagnostic signs of CI should be detectable by motor tasks. In this context, alterations in the ability of writing are considered very significant, since writing skill is the result of complex interactions between the biomechanical parts (arm, wrist, hand, etc.) and the control and memorization part of the elementary motor sequences used by each individual to produce the handwritten traces [10]. For example, in the clinical course of CI, dysgraphia occurs both during the initial phase, and in the progression of the disorder [11]. The alterations in the form and in the characteristics of handwriting can, therefore, be indicative of the onset of neurodegenerative disorders, helping physicians to make an early diagnosis. Of course, in this case, these signs are not easily visible and need a measure tools ad hoc to recognize movements and to analyze them.

Moving from these considerations, in this paper we present the results of a preliminary study in which we have considered two copy tasks of regular words and non-words, collecting the data produced by the handwriting of 99 subjects recorded by means of a graphic tablet. The rationale of our approach is to use the kinematic and pressure properties of handwritings by using some standard features proposed in the literature for testing the discriminative power of non-words to distinguish patients from healthy controls. For this aim, we considered two effective and widely used classification methods, namely Random Forest and decision trees, and a standard statistical ANOVA analysis. In particular, the paper is organized as follows: Sect. 2 presents the related work. Section 3 describes the protocol developed to collect traits of patients. Section 4 shows the structure of the dataset and feature extraction method. Section 5 displays the experiments and presents the results obtained. We conclude our paper in Sect. 6 with some future work perspectives.

2   Related Work

To date, there are many studies that investigate how variations in handwriting are prodromal indices of neurodegenerative diseases. An exhaustive review on the subject has been proposed in [5]. In [3] and in [4] instead, an experimental protocol was proposed that included various tasks which investigate the possible impaired motor cognitive functions in the AD. However, in literature, there is a difference between the type of task that seems promising to reveal some typical characteristics of the patient suffering from dementia, which seem to deviate from purely mnemonic functions. One of these types of tasks has been investigated in [1] and in [2], in which authors studied the copy tasks to support the diagnose of Alzheimer's Disease. However, the choice of the type of task has a very specific meaning. In fact, in copying tasks, unlike those of freewriting, stimuli are constantly present, and subjects can have online feedback without having recourse to memory. The cognitive impact could have consequences on the motor aspects visible from the graphic traits. In the literature, hypotheses have been made on the possibility of using words without semantic content to investigate cognitive impairment. However, many of these studies do not use online measurement tools but usually measure the final result of cognitive processing by investigating the mistakes made in terms of substitutions or inversions of letters. For example, [7] presented a literature review of the research investigating the nature of writing impairment associated with AD. They reported that in most studies words are usually categorized in regular, irregular, and non-words. Orthographically regular words have a predictable phoneme-grapheme correspondence (e.g., cat), whereas irregular words have atypical phoneme-grapheme correspondences (e.g., laugh). Non-words or pseudo-words, instead, are non-meaningful pronounceable letter strings that conform to phoneme-grapheme conversion rules and are often used to assess phonological spelling. In [9] authors proposed a writing test from dictation to 22 patients twice, with an interval of 9–12 months between the tests. They found that agraphic impairment evolved through three phases in patients with AD. The first one is a phase of mild impairment (with a few possible phonologically plausible errors). In the second phase, non-phonological spelling errors predominate, phonologically plausible errors are fewer and the errors mostly involve irregular words and non-words. The study in [8] investigated handwriting performance on a written and oral spelling task. The authors selected thirty-two words from the English language: twelve regular words, twelve irregular words and eight non-words. The study aims to find logical patterns in spelling deterioration with disease progression. The results suggested that spelling in individuals with AD was impaired relative to Healthy Control (HC). Finally, [6] used a written spelling test made up of regular words, non-words and words with unpredictable orthography. The purpose of the study was to test the cognitive deterioration from mild to moderate AD. The authors found little correlation between dysgraphia and dementia severity.

3   Acquisition Protocol

In the following subsections, the protocol designed for collecting handwriting samples and the dataset collection procedure are detailed. The 99 subjects who participated in the experiments, namely 59 CI patients and 40 healthy controls, were recruited with the support of the geriatric ward, Alzheimer unit, of the "Federico II" hospital in Naples. As concerns the recruiting criteria, we took into account clinical tests (such as PET, TAC and enzymatic analyses) and standard cognitive tests (such as MMSE). Finally, for both patients and controls, it was necessary to check whether they were on therapy or not, excluding those who used psychotropic drugs or any other drug that could influence their cognitive abilities. As previously said, the aim of the protocol is to record the dynamics of the handwriting, in order to investigate whether there are specific features that allow us to distinguish subjects affected by the above-mentioned diseases from healthy ones. The two tasks considered for this study, namely the "word" (W) task composed of two words and the "Non-Word" (NW) task composed of the other two words, require to copy four words in the appropriate box. The words chosen, as suggested in the literature, [6,9] are the two regular words of the Italian language "pane" and "mela" ("bread" and "apple" in English), and the two non-words "taganaccio" and "lonfo", i.e. nonsense words. This task aims to compare the features extracted from handwriting movements of these different types of words. The criteria according to which the structure of protocols was chosen concern:

(i) The copy tasks of NW allow us to compare the variations of the writing respect to the reorganization of the motor plan.
(ii) Tasks need to involve different graphic arrangements, e.g. words with ascenders and/or descendants, allow testing fine motor control capabilities. Indeed the regular words (W) have different descender (the "p" of the first word) and ascender traits (the "l" in the second word). The NWs propose the same structure: The first word have descender traits (the "g" in "taganaccio") and the ascender traits in the second one (the "l" and "f" in "lonfo"). Note that the NWs have to be built following the syntactic rules of language chosen.
(iii) Tasks need to involve different pen-ups that allow the analysis of air movements, which is known to be altered in the CI patients.
(iv) We have chosen to present the tasks asking the subjects to copy each word in the appropriate box. Indeed, according to the literature, the box allows the assessment of the spatial organization skills of the patient.

The two W and the two NW chosen for this study, with different ascender and descender traits, are shown in Fig. 1.

As concern the acquisition tool we have used a Graphic Tablet able to record the movements of the pen used by the examined subject. X, y, and z (pressure) coordinates are recorded for each task and saved in a txt file. The task was printed on A4 white sheet placed on the graphic tablet.

4   Structure of Dataset and Methods

The features extracted during the handwriting process have been exploited to investigate the presence of neurodegenerative diseases in the examined subjects. We used the MovAlyzer tool to process the handwritten trace, considering both on-paper and on-air traits. Their segmentation in elementary strokes was obtained by assuming as segmentation points both pen up and pen down, as well

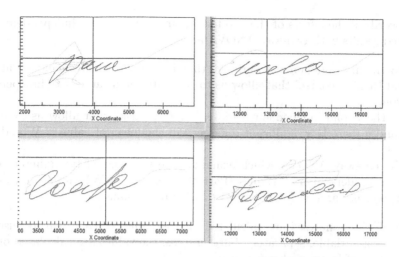

Fig. 1. Esemples of tasks. Above the regular words and below the non-words.

as the zero-crossing of the vertical velocity profile. The feature values were computed for each stroke and averaged over all the strokes relative to a single task. We also merged the W tasks between them and NW tasks, considering them as two repetitions of the same type of task. In our experiments we have included the following features for each considered type of task, separately computed for both on-paper and on-air traits: (i) Number Of Stroke; (ii) Absolute Velocity Mean; (iii) Size Mean; (iv) Loop Surface; (v) Slant Max; (vi) Horizontal Size Mean; (vii) Vertical Size mean; (viii) Total Duration; (ix) Duration Mean; (x) Absolute Size mean; (xi) Peak Vertical Velocity; (xii) Peak Vertical Acceleration; (xiii) Absolute Jerk; (xiv) Average Pen Pressure. We also included age and sex of the subject.

We have analyzed this dataset into two steps: the classification and the standard statistical analysis. For both procedures, we designed the experiments organizing the data in three groups: data obtained by extracting on paper features, data related to on-air features and data including both types of features. Thus, for each type of task, we generated three different datasets, each relative to one of the above groups and containing the samples derived for the 99 subjects. For the classification step, we used two different classification schemes, namely the Random Forest (RF) and the Decision Trees (DT) with C algorithm. For both of them, 500 iterations were performed and a 5 fold validation strategy was considered. For the statistical step, we have used the two-way ANOVA analysis and a Multiple Comparison with Holm-Sidak correction as Post Hoc analysis to understand which variables have a major effect on the results, following our 2×2 experimental design (Patient - HC; Word - NonWord). In particular, the two-factor ANOVA, used in these analyses, allows us to understand if there is a main effect on the interaction of the two factors (Label or Task), that is, in other words, if the effects of the two factors are dependent or independent. The obtained results with two-way ANOVA are:

(i) On the first independent variable, that is "Label" which includes Patients and Healthy control that allow us to compare the mean of two independent groups.
(ii) On the second independent variable, that is "task" that allow us to understand if there is a significant difference in the mean of value of W and NW tasks.
(iii) The two-way ANOVA which examines the influence of two different categorical independent variables (Patient - HC; W - NW) on one continuous dependent variable (one feature) taken into account.

All of these analyses are calculated on the values of each feature, used as a dependent variable, separately. Furthermore, the values are calculated on the three groups of features shown above.

5   Experimental Results

As concerns the classification, in Fig. 2 we summarize the values of Accuracy for each group of tasks. The first column reports the type of task considered, the second one the classifier employed, while the following columns report, for each task, the value of Accuracy using all features, on-paper features and on-air features respectively.

From this table, we can point out that: firstly, in the large majority of cases, for each task the value of the Accuracy is over 80.00%, reaching the best value in the second group of tasks (NW) equal to 85.35 %. Secondly, we can observe that, in the same condition of considered feature, emerges a better classification using the RF classifier compared to DT. This is easily justifiable considering that Random Forest, unlike DT, is an ensemble of classifiers. However, as reported in the last row, the best-obtained result occurs with NW task. Indeed the classification accuracy of the second task is almost always 5% higher than the first type of tasks. This means that copying a word with no meaning can be more useful to diagnose CI than copying a regular word and that the same features better predict CI in NW task than W task.

In the Fig. 3 we show, instead, the value of standard statistical analysis, with two-way ANOVA, using R tool. In this table, we indicate, for each group of tasks, the statistical significance results of each feature taken into account. We report the value of the interaction of rows and columns (Label and Tasks) separately, and the value of their interaction (in grey), according to the translation criteria shown as footnote. Empty cells indicate that no statistical significance emerged for these factors.

We can claim that there is a high significance value for almost all features on the two factors separately. In particular:

Task	Classifier	Accuracy		
		All	OnPap	OnAir
Word	J48	78.79	81.52	78.79
	RF	82.32	83.15	82.83
Non-Word	J48	84.85	82.06	80.81
	RF	**84.85**	**83.69**	**85.35**

Fig. 2. Accuracy of Words and Non-Words tasks

(i) As regards all features we can note that there is an interaction on Number of Stroke, Absolute Velocity Mean, Total Duration and Jerk. As reported in Fig. 4 and in detail in Fig. 7:
   (i) the Variation of Number of Stroke is significant for each combination, except for W:P vs NW:HC. Patients produce almost double Number of Stroke of HC both for W and NW condition. Furthermore, HCs produce fewer strokes in NW than Patients in W condition.
   (ii) the Duration Tot is doubled both for P vs HC and W vs NW. But it is worth noticing that HCs employ less time to complete NW task than P to produce W task.
   (iii) for the Absolute Velocity Mean and for Jerk the value of HCs increases going from W to NW task but conversely, it decreases for patients.
(ii) As regards on-air features we can note that there is an interaction on Number of Stroke, Absolute Velocity Mean, Total Duration, Jerk and Loop Surface Mean. As reported in Fig. 5 and in detail in Fig. 7:
   (i) the variation of the Number of Stroke on air is even more evident than in all features. Patients produce almost double Number of Stroke of HCs both for W and NW condition. Moreover, HCs produce more stroke of NW than Patients in W condition.
   (ii) the Duration Tot is doubled both for P vs HC and W vs NW.
   (iii) for the Absolute Velocity Mean and for the Jerk, the values show the same relation compared to all feature but, in this case, the difference between HC and Patient in W task in Abs Velocity Mean is irrelevant.

(iii) Finally, as regards on paper features we can note that there is an interaction only on Slant and Peak Acceleration Mean. As reported in Fig. 6 and in detail in Fig. 7:

(i) The variation of the Slant is significant for almost all comparisons and HC in NW is less than Slant in P in W condition.

(ii) Peak Acceleration Mean shows an opposite trend between HCs and Patients in the comparison of W and NW tasks.

The second experimental setting has shown that some features can be particularly discriminative of the disease. For example, Absolute Velocity Mean and Jerk on all features and on-air features show an inverted trend between HCs and Ps in the passage from W tasks to NW tasks.

FEATURE	ALL		ON AIR		ON PAPER	
	LABEL	TASK	LABEL	TASK	LABEL	TASK
NumOfStroke *(interaction: ALL `*`, ON AIR `*`)*	***	***	***	***	***	****
AbsoluteVelocityMean *(interaction: ALL `.`, ON AIR `*`)*	***		*		***	
SizeMean					***	
LoopSurfaceMean *(interaction: ON AIR `*`)*	***				***	.
SlantMax *(interaction: ON PAPER TASK `.`)*	***	***	***	***	***	**
HorizontalSizeMean	***	***	**		***	***
VerticalSizeMean			**		**	**
DurationTot *(interaction: ALL `*`, ON AIR `**`)*	***	***	***	***	***	***
DurationMean						**
AbsoluteSizeMean	***		**			
PeakVerticalVelocityMean				**		
PeakVerticalAccelerationMean *(interaction: ON PAPER `.`; ALL TASK `.`)*						
JerkMean *(interaction: ALL `*`, ON AIR `**`; ON AIR LABEL `.`)*					***	
PenPressure Mean	***	**			***	

Signif. codes: 0 '***' 0.001 '**' 0.01 '*' 0.05 '.' 0.1 ' ' 1

Fig. 3. Statistical significance with ANOVA

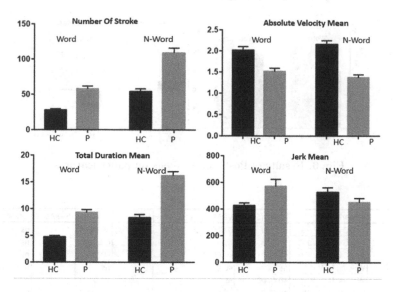

Fig. 4. Results of Post Hoc analysis on all features

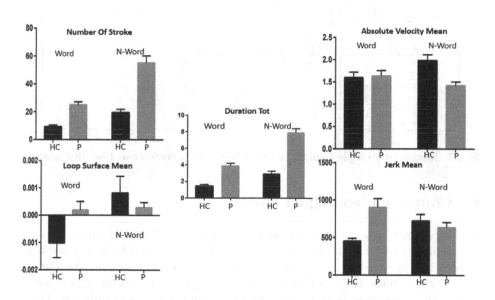

Fig. 5. Results of Post Hoc analysis on air features

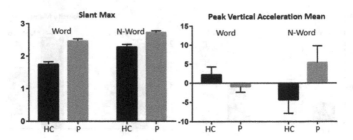

Fig. 6. Results of Post Hoc analysis on paper features

	Comparison	NumOfStroke	VelMean	Loop	Slant	DurTot	PeakAccel	JerkMean
ALL	W:HC vs. W:P	0.0009	0.0001			<0.0001		ns
	W:HC vs. N W:HC	0.0118	ns			0.0024		ns
	W:HC vs. N W:P	<0.0001	<0.0001			<0.0001		ns
	W:P vs. N W:HC	ns	<0.0001			ns		ns
	W:P vs. N W:P	<0.0001	ns			<0.0001		ns
	N W:HC vs. N W:P	<0.0001	<0.0001			<0.0001		ns
OnAir	W:HC vs. W:P	0.0166	ns	ns		0.0005		0.0090
	W:HC vs. N W:HC	ns	ns	0.0232		ns		ns
	W:HC vs. N W:P	<0.0001	ns	ns		<0.0001		ns
	W:P vs. N W:HC	ns	ns	ns		ns		ns
	W:P vs. N W:P	<0.0001	ns	ns		<0.0001		ns
	N W:HC vs. N W:P	<0.0001	0.0084	ns		<0.0001		ns
OnPap	W:HC vs. W:P				<0.0001		ns	
	W:HC vs. N W:HC				<0.0001		ns	
	W:HC vs. N W:P				<0.0001		ns	
	W:P vs. N W:HC				ns		ns	
	W:P vs. N W:P				0.0042		ns	
	N W:HC vs. N W:P				<0.0001		ns	

Fig. 7. Results of Multiple Comparison with Holm-Sidal correction- Post Hoc analysis

6   Conclusion and Open Issues

In this paper, we presented a novel solution to diagnose Cognitive impairment by means of only two writing tasks which included a copy of regular words and non-words. The preliminary results obtained are very encouraging and the work is in progress to increase general performance. As appeared from the experimental results there are some features particularly discriminative of the disease. Then, the next steps to be taken will include a classification experiment using just these features. We could also include a feature selection approach to improve overall performances.

References

1. Cilia, N.D., De Stefano, C., Fontanella, F., Molinara, M., Scotto di Freca, A.: Handwriting analysis to support alzheimer's disease diagnosis: a preliminary study. In: Vento, M., Percannella, G. (eds.) CAIP 2019. LNCS, vol. 11679, pp. 143–151. Springer, Cham (2019). https://doi.org/10.1007/978-3-030-29891-3_13
2. Cilia, N.D., De Stefano, C., Fontanella, F., Molinara, M., Scotto di Freca, A.: Using handwriting features to characterize cognitive impairment. In: Ricci, E., Rota Bulò, S., Snoek, C., Lanz, O., Messelodi, S., Sebe, N. (eds.) ICIAP 2019. LNCS, vol. 11752, pp. 683–693. Springer, Cham (2019). https://doi.org/10.1007/978-3-030-30645-8_62
3. Cilia, N.D., De Stefano, C., Fontanella, F., Scotto di Freca, A.: An experimental protocol to support cognitive impairment diagnosis by using handwriting analysis. Procedia Comput. Sci. **141**, 466–471 (2018)
4. Cilia, N.D., De Stefano, C., Fontanella, F., Di Freca, A.S.: Using genetic algorithms for the prediction of cognitive impairments. In: Castillo, P.A., Jiménez Laredo, J.L., Fernández de Vega, F. (eds.) EvoApplications 2020. LNCS, vol. 12104, pp. 479–493. Springer, Cham (2020). https://doi.org/10.1007/978-3-030-43722-0_31
5. De Stefano, C., Fontanella, F., Impedovo, D., Pirlo, G., Scotto di Freca, A.: Handwriting analysis to support neurodegenerative diseases diagnosis: a review. Pattern Recogn. Lett. **121**, 37–45 (2018)
6. Luzzatti, C., Laiacona, M., Agazzi, D.: Multiple patterns of writing disorders in dementia of the alzheimer-type and their evolution. Neuropsychologia **41**(7), 759–772 (2003)
7. Neils-Strunjas, J., Groves-Wright, K., Mashima, P., Harnish, S.: Dysgraphia in Alzheimer's disease: a review for clinical and research purposes. J. Speech Lang. Hear. Res. **49**(6), 1313–1330 (2006)
8. Pestell, S., Shanks, M.F., Warrington, J., Venneri, A.: Quality of spelling breakdown in alzheimer's disease is independent of disease progression. J. Clin. Exp. Neuropsychol. **22**(5), 599–612 (2000)
9. Platel, H., et al.: Characteristics and evolution of writing impairment in alzheimer's disease. Neuropsychologia **31**(11), 1147–1158 (1993)
10. Tseng, M.H., Cermak, S.A.: The influence of ergonomic factors and perceptual-motor abilities on handwriting performance. Am. J. Occup. Ther. **47**(10), 919–926 (1993)
11. Yan, J.H., Rountree, S., Massman, P., Doody, R.S., Li, H.: Alzheimer's disease and mild cognitive impairment deteriorate fine movement control. J. Psychiatr. Res. **42**(14), 1203–1212 (2008)

Modeling the Coordination of a Multiple Robots Using Nature Inspired Approaches

Mauro Tropea[✉], Nunzia Palmieri, and Floriano De Rango

DIMES Department, University of Calabria, P.Bucci 39/c, 87036 Rende (CS), Italy
{mtropea,npalmieri,derango}@dimes.unical.it

Abstract. The work focuses on the problem of multiple robots coordination in search and rescue mission. In particular, decentralized swarm techniques, that use mechanisms based on Swarm Intelligence, are presented. Essentially, two approaches are compared. The first uses a one-hop communication mechanism to spread locally the information among the robots and a modified Firefly meta-heuristics is proposed. The second approach, is based on a multi-hop communication mechanism based on Ant Colony Optimization. We have conducted experiments for evaluating what is the best approach to use considering different parameters of the system.

Keywords: Multi robots coordination · Nature inspired techniques · Swarm Intelligence

1 Introduction

Over the past decade, the field of distributed robotics system has been investigated actively involving multiple robots coordination strategy design. The field has grown dramatically, with a wider variety of topics being addressed, where multi-robot systems can often deal with complex tasks, that cannot be accomplished by an individual robot. However, the use of multiple robots poses new challenges; indeed, the robots must communicate and coordinate in such a way that some predefined global tasks can be achieved more efficiently. Swarm robotics is a new approach to the coordination of multi-robot systems which consists of large numbers of mostly simple physical robots. It gets inspiration from Swarm Intelligence (SI) to model the behavior of the robots.

SI-based algorithms are among the most popular and widely used. There are many reasons for such popularity, one of the these is that SI-based algorithms usually share information among multiple agents, so that self-organization,co-evolution and learning during iterations may help to provide the high efficiency of most SI-based algorithms. Another reason is that multiple agent can be parallelized easily so that large-scale optimization becomes more practical from the implementation point of view [1–4]. Recently, SI-based algorithms have applied also in search and rescue operations to coordinate teams of robots [5].

F. Cicirelli et al. (Eds.): WIVACE 2019, CCIS 1200, pp. 124–133, 2020.
https://doi.org/10.1007/978-3-030-45016-8_13

Some years ago, a novel technique called Firefly algorithm has been proposed to realize a multi-modal optimization for hard-decision problem [6]. Its basic version has been modified and extended to fit search and rescue operations in the field of mobile robots such as presented in [7,8]. Moreover, a distributed wireless protocol has been also proposed in [9] to speed up the execution time of multiple tasks such as exploring and disarming task. In the past years also the exploration of unknown area through mobile robots has been considered using Ant Colony Optimization (ACO) inspired approaches [10,11]. Some works considered hazard environment with limited communication capability such as in [12] andother works have been proposed for routing based on ACO [13,14]. However, few of the proposed strategies in exploration and target finding, considered energy consumption as evaluation metrics to estimate the effectiveness of the proposed approach. Some distributed protocols applied in the context of mobile ad-hoc networks where the energy issue is considered are presented in [15–18]. Moreover, some multi-objective formulations for the path finding and bio-inspired solutions have been proposed. Recent works considered also the SI algorithms applied in the decision-making process and security such as presented respectively in [19,20] and in routing and coordination issues such as in [21–26].

Main contributions of the paper are listed below:

- A math formulation for the search and rescue operations to mine disarming in an unknown area is presented. This formulation considers some constraints in the problem such as minimum number of robots to perform the disarming task and the discovery of all cells in the unknown area.
- Two bio-inspired techniques have been introduced evaluating their performance in terms of scalability for increasing targets and number of robots.The first one is inspired by FireFly (FF) algorithm and it makes use of local communication; the second one is inspired by ACO and it is based on a distributed multi-hop wireless protocol to disseminate data.
- Two coordination techniques, FireFly based Team Strategy for Robot Recruitment (FTS-RR) and Ant based Task Robots Coordination (ATRC) are compared in terms of total time steps for completing the mission varying the number of targets and the grid area size and in terms of average energy for a robot varying the robots' coalitions.

2   Problem Description

In this work, we consider a set R of homogeneous robots working in a discrete domain [7–9]. Each robot has limited sensing capabilities. The communication range of the robots is assumed to be limited, and a robot can reach another robot by a sequence of communication links. Furthermore, the robots have a limited computational power, so their cooperative strategies cannot involve complex planning or negotiations. In the area, a certain number of different targets are scattered randomly. Each target z requires a certain amount of robot (R_{min}) to be handled. It is assumed that there are no a priori knowledge about the targets such as locations and numbers. When a robot detects a target, it is assumed that the sensing information is perfect. Since a single robot does not have enough resources to handle the target, a coalition of robots may need to be formed to jointly handle the target z safely. In the Fig. 1 a representation of the considered environment and an example of robots' coalitions are shown. In the Table 1 the description of some variables of the system used in the mathematical formulation is provided. The problem, accounting both the exploration time and the coordination time, can be mathematically stated as follows:

Table 1. Variable of the system.

Variable	Description
A	grid map and $A \subset \Re^2$
R	set of robots
N^R	number of robots $|N^R| = R$
R_{min}	number of robots needed to deal with a target
T	set of targets
N^T	number of targets, $N^T = |T|$
v_{xy}^k	$\begin{cases} 1 \; if \; the \; robot \; k \; visits \; the \; cell \; of \; coordinates \; (x,y) \\ 0 \; otherwise \end{cases}$
T_e	Time to visit a cell
$T_{coord,z}^k$	Time to coordinate a target z
u_z^k	$\begin{cases} 1 \; if \; the \; robot \; k \; is \; involved \; in \; the \; target \; z \\ 0 \; otherwise \end{cases}$
E_m^k	The energy consumed by the robot k for moving
E_{tx}^k	The energy consumed by the robot k to transmit a packet
E_{rx}^k	The energy consumed by the robot k to receive a packet
E_d^k	The energy consumed by the robot k to deal with a target
E_{coord}^k	Energy consumed by the robot k for coordination

$$minimize \sum_{k=1}^{N^R} \sum_{x=1}^{m} \sum_{y=1}^{n} T_e v_{xy}^k + \sum_{k=1}^{N^R} \sum_{z=1}^{N^T} T_{coord,z}^k u_z^k \qquad (1)$$

subject to

$$\sum_{k=1}^{N^R} v_{xy}^k \geq 1 \quad \forall (x,y) \in A \qquad (2)$$

$$\sum_{k=1}^{N^R} u_z^k = R_{min} \quad \forall z \in T \qquad (3)$$

$$v_{xy}^k \in 0,1 \quad \forall(x,y) \in A, \ k \in R \tag{4}$$

$$u_z^k \in 0,1 \quad \forall z \in A, \ k \in R \tag{5}$$

$$T_e, T_{coord,z}^k, E_m^k, E_{coord}^k \in R, \ \forall z \in T, \ k \in R \tag{6}$$

Where $E_{coord}^k = \sum_{z=1}^{N^T}(E_{tx}^k + E_{rx}^k + E_d^k) \cdot u_z^k$, is the the energy consumed for the coordination task by the robot k that is involved in the target.

Equation 2 and Eq. 3 are respectively the constraints that assure that all cells need to be visited (Eq. 2) at least by one robot and the minimum number of robots requested to complete the demining task R_{min}. In the considered problem is evaluated the energy consumption of robots on the basis of the following equation:

$$E_{tot} = \sum_{k=1}^{N^R} E_m^k + \sum_{k=1}^{N^R} E_{coord}^k \tag{7}$$

Equation 7 is useful to compute the total energy consumed by robots during the movement and in the specific task of demining. This means that the contribution of energy accounts for the movement and the search operations whereas the second contribution considers that communication cost (transmission and reception of recruiting packets) and the demining task when the robots reaches the target location.

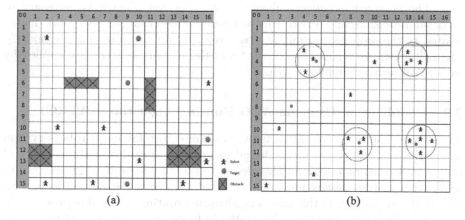

Fig. 1. (a) A representation of the considered environment. (b) Local coalitions of robots formed through the recruitment's processes.

3   Multi-robots Architecture Design

We suppose that the robots explore the area using an algorithm such as presented in [7–9]. The work treats only the problem of recruiting the needed robots in targets locations using only local spreading of the information about the detected targets. A network architecture is created once a robot detects a target in the area and from this point that initiates communication with neighbour to neighbour. One approach uses only one hop communication and a Firefly based algorithm is used to coordinate the robots. It is called FireFly based Team Strategy for Robot Recruitment (FTS-RR). The other approach regards the development of a protocol to coordinate the team. The strategy is called Ant based Task Robots Coordination (ATRC) protocol. In this case the robots exchange simple information to avoid the redundancy in reaching the targets location. The communication is multi-hop and higher number of robots can be reached to be recruited.

3.1   Coordination Using a Firefly Algorithm (FTS-RR)

Concerning the considered problem, the robots that have detected a target, start to behave like a Firefly sending out help requests to its neighbourhood. When a robot k receives this request and it decides to contribute in the disarming process, it stores the request in its list. If the list contains more requests, it must choose which target it will disarm. Using the relative position information of the found targets, the robot derives the distance between it and the coordinators and then it uses this metric to choose the best target, that is usually the closer. In this case, no forwarding of the packet is done and the communication is one hop. The approach provides a flexible way to decide when it is necessary to reconsider decisions and how to choose among different targets. For more details related to the proposed technique please refer to our previous contribution [7]. Figure 2 illustrates a simplified flowchart of the ACO-based strategy applied by each robot for the Firefly algorithm.

3.2   Coordination Using an ACO Routing Algorithms (ATRC)

The second approach presents a network architecture for multi robots system where the information about the found targets can spread over the network of robots in a multi-hop fashion. The idea is to use an ad-hoc routing protocol to report the detected targets and the robots that want to help in disarming process over a MANET. Also, in this case a bio-inspired routing protocol is proposed in order to reduce the communication traffic in terms of packets and allows at the same time a self-adaptive behaviour of the robots. More specifically, the protocol takes inspiration from the ability of certain types of ants in nature to find the shortest path between their nest and a food source through a distributed process based on stigmergic communication. ACO based routing algorithms can usually set multiple paths, over which data packets can be forwarded probabilistically

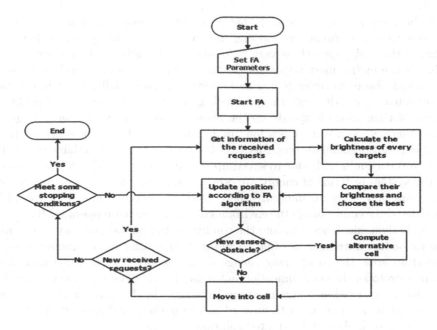

Fig. 2. The flow chart of Firefly algorithm.

like ants. This can result in throughput optimization, automatic data load balancing, and increased robustness to failures. The network of robots is created when one or more robots find a target. More specifically, the robot that has detected a target sends announcement messages that are forwarded by the other robots so that the information about the target can spread among the swarm. The messages that a robot can send or receive are: Hello packets, Requiring Task Forward Ant (RT-FANT) that is a packet sent by the robot that has detected a target to know how many robots are available to treat the target; Requiring Task Backward Ant (RT-BANT): it is a packet that a robot sends as response to a RT-FANT; Recruitment Fant (R-FANT): it is a packet sent by a coordinator, to the link from which came the higher number of RT-BANT responses; this link has higher recruitment probability; Recruitment Bant (R-BANT): it is a packet sent by a robot in response to a positive recruitment by a coordinator [9].

4 Simulations

In this work we want to compare the two approaches trying to understand what is the best one. Performance metrics considered for the simulation are: Total Time steps to complete the mission and the Average energy consumed by a robot for accomplishing the mission. We suppose that for treating a target it is required the work of 3 robots together. The convergence time and the energy was averaged over 100 independent simulation runs. Regarding the parameter of the techniques please refer to [7]. A computational study and extensive simulations have been carried out to assess the behaviour of the proposed approaches and to analyse their performance by varying the parameters of the problem. Establish what is the best approach, is very hard since it depends on the context and on what is the metric most important. Figure 3 shows the main simulation results described above, in order to try to understand, potentially, what is the best approach to use. Although the protocol, generally, can offer more benefits in terms of time, since it speeds up the mission, the consumed energy is grater since there is more communication among the team. The best approach to be used depends on many factors. Firstly, if the time is a critical variable thus using the protocol could be better to speed up the mission. Secondly, if the resources of the system in terms of energy are crucial using only local interaction among the team may allow to minimize the consumed energy. Thirdly, it should be considered the conditions of the environment where the team operates. If the area is highly dynamic, hazardous and the conditions to maintain the network among the robots are unreliable, it could be suitable adapt one hop communication. In uncertain area, the robots may change decisions anytime. In these situations, using a protocol the communication in terms of packets could increase and thus lead to an overhead of communication. However, the designed protocol is based on probabilistic mechanism to forward of the packets and make decision, so it can offer a scalable and distributed solution.

a)	20 robots			30 robots			40 robots		
	1 target	3 targets	6 targets	1 target	3 targets	6 targets	1 target	3 targets	6 targets
FTS-RR	103	178	173	91	109	106	74	95	75
ATRC	96	118	130	86	98	105	75	80	85

b)	20 robots			30 robots			40 robots		
	Grid map 30x30	Grid map 50x50	Grid map 60x60	Grid map 30x30	Grid map 50x50	Grid map 60x60	Grid map 30x30	Grid map 50x50	Grid map 60x60
FTS-RR	178	294	434	109	171	284	95	131	203
ATRC	105	230	340	99	190	280	85	153	210

c)	5 targets			7 targets			10 targets		
	Grid map 30x30	Grid map 50x50	Grid map 60x60	Grid map 30x30	Grid map 50x50	Grid map 60x60	Grid map 30x30	Grid map 50x50	Grid map 60x60
FTS-RR	455	689	898	499	729	950	464	805	993
ATRC	625	958	1228	700	999	1155	725	1050	1305

d)	5 targets			7 targets			10 targets		
	Grid map 30x30	Grid map 50x50	Grid map 60x60	Grid map 30x30	Grid map 50x50	Grid map 60x60	Grid map 30x30	Grid map 50x50	Grid map 60x60
FTS-RR	333	581	676	395	689	676	435	791	887
ATRC	510	674	839	540	697	839	557	719	994

Fig. 3. Comparison of FTS-RR and ATRC: a) Total Time Steps to complete the mission varying the number of targets; b) Total Time Steps to complete the mission considering varying the grid area size; c) Average Energy for a Robot in term of units of charge considering a coalition of 20 robots; d) Average Energy for a Robot in term of units of charge considering a coalition of 30 robots.

5 Conclusions

A brief comparison between two bio-inspired strategy to perform demining task has been proposed. The first approach (FTS-RR) is based on a local communication and it is inspired to FireFly (FF) algorithm whereas the second approach (ATRC) make use of exploring ANTs to know the network topology in order to speed up the completion of demining task. Both techniques can be useful in search and recruiting tasks. However, by simulations it is possible to see as the FF algorithm seems to be more scalable when the convergence time is not so important in comparison with the energy consumption. This technique seems to perform better in condition of higher number of robots or higher number of targets especially in terms of energy consumption. On the other hand, ATRC seems to be more performing in terms of task execution because the knowledge of the topology and robots disposition can speed up the recruiting task reducing the overall time to search and recruit. However, simulations show the degradation of ATRC in terms of energy consumption because a protocol and more communication among robots become necessary.

References

1. Khamis, A., Hussein, A., Elmogy, A.: Multi-robot task allocation: a review of the state-of-the-Art. In: Koubâa, A., Martínez-de Dios, J.R. (eds.) Cooperative Robots and Sensor Networks 2015. SCI, vol. 604, pp. 31–51. Springer, Cham (2015). https://doi.org/10.1007/978-3-319-18299-5_2
2. Bayındır, L.: A review of swarm robotics tasks. Neurocomputing **172**, 292–321 (2016)
3. Bakhshipour, M., Ghadi, M.J., Namdari, F.: Swarm robotics search & rescue: a novel artificial intelligence-inspired optimization approach. Appl. Soft Comput. **57**, 708–726 (2017)
4. Brambilla, M., Ferrante, E., Birattari, M., Dorigo, M.: Swarm robotics: a review from the swarm engineering perspective. Swarm Intell. **7**(1), 1–41 (2013)
5. Senanayake, M., Senthooran, I., Barca, J.C., Chung, H., Kamruzzaman, J., Murshed, M.: Search and tracking algorithms for swarms of robots: a survey. Robot. Autonom. Syst. **75**, 422–434 (2016)
6. Yang, X.-S.: Firefly algorithms for multimodal optimization. In: Watanabe, O., Zeugmann, T. (eds.) SAGA 2009. LNCS, vol. 5792, pp. 169–178. Springer, Heidelberg (2009). https://doi.org/10.1007/978-3-642-04944-6_14
7. Palmieri, N., Yang, X.-S., De Rango, F., Marano, S.: Comparison of bio-inspired algorithms applied to the coordination of mobile robots considering the energy consumption. Neural Comput. Appl. **31**(1), 263–286 (2017). https://doi.org/10.1007/s00521-017-2998-4
8. Palmieri, N., Yang, X.S., De Rango, F., Santamaria, A.F.: Self-adaptive decision-making mechanisms to balance the execution of multiple tasks for a multi-robots team. Neurocomputing **306**, 17–36 (2018)

9. De Rango, F., Palmieri, N., Yang, X.-S., Marano, S.: Swarm robotics in wireless distributed protocol design for coordinating robots involved in cooperative tasks. Soft Comput. **22**(13), 4251–4266 (2017). https://doi.org/10.1007/s00500-017-2819-9

10. Schroeder, A., Ramakrishnan, S., Kumar, M., Trease, B.: Efficient spatial coverage by a robot swarm based on an ant foraging model and the Lévy distribution. Swarm Intell. **11**(1), 39–69 (2017)

11. Oh, H., Shirazi, A.R., Sun, C., Jin, Y.: Bio-inspired self-organising multi-robot pattern formation: a review. Robot. Autonom. Syst. **91**, 83–100 (2017)

12. Gregory, J., et al.: Application of multi-robot systems to disaster-relief scenarios with limited communication. In: Wettergreen, D.S., Barfoot, T.D. (eds.) Field and Service Robotics. STAR, vol. 113, pp. 639–653. Springer, Cham (2016). https://doi.org/10.1007/978-3-319-27702-8_42

13. De Rango, F., Tropea, M., Provato, A., Santamaria, A.F., Marano, S.: Minimum hop count and load balancing metrics based on ant behavior over HAP mesh. In: 2008 IEEE Global Telecommunications Conference on IEEE GLOBECOM 2008, pp. 1–6. IEEE(2008)

14. De Rango, F., Tropea, M., Provato, A., Santamaria, A.F., Marano, S.: Multiple metrics aware ant routing over HAP mesh networks. In: 2008 Canadian Conference on Electrical and Computer Engineering, pp. 001675–001678. IEEE, May 2008

15. Singh, G., Kumar, N., Verma, A.K.: Ant colony algorithms in MANETs: a review. J. Network Comput. Appl. **35**(6), 1964–1972 (2012)

16. Fotino, M., De Rango, F.: Energy issues and energy aware routing in wireless ad hoc networks, pp. 281–296. INTECH Open Access Publisher (2011)

17. De Rango, F., Lonetti, P., Marano, S.: MEA-DSR: a multipath energy-aware routing protocol for wireless Ad Hoc networks. In: Cuenca, P., Guerrero, C., Puigjaner, R., Serra, B. (eds.) Advances in Ad Hoc Networking. IIFIP, vol. 265, pp. 215–225. Springer, Boston, MA (2008). https://doi.org/10.1007/978-0-387-09490-8_19

18. De Rango, F., Tropea, M.: Energy saving and load balancing in wireless Ad Hoc networks through ant-based routing. In: 2009 International Symposium on Performance Evaluation of Computer & Telecommunication Systems, vol. 41, pp. 117–124. IEEE, July 2009

19. Rizk, Y., Awad, M., Tunstel, E.W.: Decision making in multiagent systems: a survey. IEEE Trans. Cogn. Dev. Syst. **10**(3), 514–529 (2018)

20. Strobel, V., Castelló Ferrer, E., Dorigo, M.: Managing byzantine robots via blockchain technology in a swarm robotics collective decision making scenario. In: Proceedings of the 17th International Conference on Autonomous Agents and MultiAgent Systems, pp. 541–549. International Foundation for Autonomous Agents and Multiagent Systems, July 2018

21. De Rango, F., Tropea, M.: Swarm intelligence based energy saving and load balancing in wireless Ad Hoc networks. In: Proceedings of the 2009 Workshop on Bio-inspired Algorithms for Distributed Systems, pp. 77–84. ACM (2009)

22. De Rango, F., Potrino, G., Tropea, M., Santamaria, A.F., Fazio, P.: Scalable and ligthway bio-inspired coordination protocol for FANET in precision agriculture applications. Comput. Electric. Eng. **74**, 305–318 (2019)

23. De Rango, F., Palmieri, N., Tropea, M., Potrino, G.: UAVs team and its application in agriculture: a simulation environment. SIMULTECH **2017**, 374–379 (2017)

24. De Rango, F., Potrino, G., Tropea, M., Santamaria, A.F., Palmieri, N.: Simulation, modeling and technologies for drones coordination techniques in precision agriculture. In: Obaidat, M.S., Ören, T., Rango, F.D. (eds.) SIMULTECH 2017. AISC, vol. 873, pp. 77–101. Springer, Cham (2019). https://doi.org/10.1007/978-3-030-01470-4_5

25. Tropea, M., Santamaria, A.F., Potrino, G., De Rango, F.: Bio-inspired recruiting protocol for FANET in precision agriculture domains: pheromone parameters tuning. In: 2019 Wireless Days (WD), pp. 1–6. IEEE, April 2019
26. De Rango, F., Tropea, M., Fazio, P.: Bio-inspired routing over FANET in emergency situations to support multimedia traffic. In: Proceedings of the ACM Mobi-Hoc Workshop on Innovative Aerial Communication Solutions for FIrst REsponders Network in Emergency Scenarios, pp. 12–17, July 2019

Nestedness Temperature in the *Agent-Artifact* Space: Emergence of Hierarchical Order in the 2000–2014 Photonics Techno-Economic Complex System

Riccardo Righi[✉][iD], Sofia Samoili[✉][iD], Miguel Vazquez-Prada Baillet,
Montserrat Lopez-Cobo[iD], Melisande Cardona[iD], and Giuditta De Prato[iD]

European Commission, Joint Research Centre (JRC),
Unit B6 - Digital Economy, Seville, Spain
{riccardo.righi,sofia.samoili,miguel.vazquez-prada-baillet,
montserrat.lopez-cobo,melisande.cardona,giuditta.de-prato}@ec.europa.eu

Abstract. In this work we represent a techno-economic complex system based on the *agent-artifact* space theoretical model. The objective is to structure a methodology to statistically investigate the presence of hierarchical order, as an emerging property of this system. To analyse the agent-artifact space, two statistical methodologies are initially employed. The first is a community detection method, employed with the objective to detect groups of agents that are likely to intensively exchange information within the considered complex system. The second is a natural language processing method, the LDA topic model, employed with the objective of identifying types of artifacts as technological subdomains through textual information that describes the activities of agents. After this initial part, we address the investigation of the structure of the *agent-artifact* space by estimating the involvement of each community in the detected topics. This is effectuated by means of a statistic that considers the information flow percentage of agents, the fractional count of activities, and the probability of agents' activities to belong to topics. We then estimate the hierarchical order of the topics' distribution in communities, by computing its nestedness temperature, which is adopted by studies on ecological systems. This statistic's significance is finally evaluated with z-scores based on homogeneous systems. The case study is a system consisted of economic agents (e.g. firms, universities, governmental institutions) patenting in the technological domain of photonics. The analysis is effectuated over five time spans in the period 2000–2014. The observed values of nestedness temperature are proved statistically significant, which suggests that hierarchical order is an emerging property of the *agent-artifact* space.

European Commission—The views expressed are purely those of the author and may not in any circumstances be regarded as stating an official position of the European Commission.

F. Cicirelli et al. (Eds.): WIVACE 2019, CCIS 1200, pp. 134–144, 2020.
https://doi.org/10.1007/978-3-030-45016-8_14

1 Introduction

The present work aims to investigate the emergent presence of physical order in an observed techno-economic complex system. Initially, two parallel methodologies are implemented with the respective objectives of (i) detecting group of agents that are likely to interact among themselves, and (ii) identifying the technological thematic subdomains that emerge under the initially considered technology, i.e. 'photonics'. As described in the relevant literature regarding the conceptualization of the *agent-artifact space* [19–21], two elements have to be considered when analyzing evolutionary dynamics regarding innovation and technological developments. The first element is the presence of interactive meso-structures of *agents*, which in this work are investigated as communities of agents (i.e. groups) intensively interacting among themselves. The second element is the presence of different kinds of *artifacts* belonging to the same technological domain. These are investigated in this study as thematic topics (i.e. technological subdomains). From a methodological point of view, the first part of the analysis is developed based on a community detection over a multilayer network (MLN), while the second analysis is developed based on an unsupervised natural language processing method, namely topic modeling.

After the detection of communities and topics, the distribution of topics in communities is analyzed with an approach based on ecology [3,27]. In order to look for the existence of any hierarchical structure, we investigate nestedness. In ecological systems, one of the interpretations of the nestedness's presence is that: (i) species that are located in the largest number of different habitats, are the only ones located in the habitat presenting the minimum diversity, and (ii) species that are located in few habitats, are located in habitats that are populated by a large diversity of species. When either of these cases occurs, the system is hierarchically ordered. In order to measure this order, the nestedness is measured as the temperature of the matrix describing the presence of species in habitats [3], where a lower temperature indicates a more nested system. Finally, to assess the presence of order in the considered *agent-artifact space*, i.e. a complex system of economic institutions patenting in photonics from 2000 to 2014, we refer to the previously introduced concepts. In the matrix describing the involvement of communities of agents in the technological subdomains, we compute the nestedness temperature. Then its statistical significance is evaluated with computed homogeneous systems. The obtained results prove that the detected levels of nestedness are statistically significant in the considered case study.

The originality of this work is the development of a method to analyze techno-economic complex systems towards two directions. The first is the use of the conceptualization of the *agent-artifact* space in order to structure the analysis of a techno-economic complex system. The second is the investigation of its emerging properties and, more specifically for this work, the presence of physical order, which is here measured in terms of nestedness. Therefore, we direct the investigation of agents' interactive groups (the communities) and of the existing types of artifacts (the topics), towards the measurement of their intertwining.

The economic implications of the implemented analysis and a prediction model based on the obtained results, are not developed in this work. Both these will be the subject of subsequent studies.

In Sect. 2, we introduce the methodology developed to detect communities of agents and technological topics. In Sect. 3 we refer to nestedness temperature to assess the involvement of the communities in the topics, and we measure its significance. Finally, in Sect. 4 we implement the methodology in a real observed system and we discuss the obtained results.

2   A Statistical Approach for the Representation of the *Agent-Artifact* Space

The informational basis needed for the implementation of the proposed methodology is a bipartite network made of economic agents (e.g. companies, universities, research centers, local government institutions) participating in economic activities related to innovation processes (in this work, for instance, we consider patents). Apart from the aforementioned network structure, the requested information is dual: the geographic location of the agents, and the textual information describing the economic activities in which they are involved. On this basis, two methodologies are implemented in parallel [25]. These are a (i) community detection based on a MLN structure, and (ii) a topic modelling based on the textual technological information.

The first, i.e. the community detection, aims to detect groups of agents that are likely to exchange information, due to proximity determined by interactions in multiple dimensions. The second, i.e. the topic modelling, aims to identify technological subdomains, in order to categorize the innovative economic activities based on their content. These two analyses are selected according to the relevant literature regarding the conceptualization of the *agent-artifact space* [19–21], in which innovation processes are explained to be the results of a series of interactions that (i) occur among the agents involved in the system, and (ii) whose core objective is the discussion about the content of a specific technological artifact. In order, to pertinently develop the methodology with respect to the considered theoretical approach, we investigate how agents interacted, and about what they interacted. In the rest of the section, we present the two parallel analyses that constitute the first part of our work.

A MLN Representing a Multi-dimensional Space of Interactions. The initial bipartite network of agents and activities, conceptually *agents* and *artifacts*, is transformed in its one-mode projection. The result is a one-mode network in which agents are connected among themselves based on the common activities they performed together. Then, two additional networks are generated. In the first, the same agents of the aforementioned network are connected only when belonging to the same geographical area, which in this work is described as sub-regions. In the second, the same agents are connected on the basis of the use of similar keywords in the patents they developed. These three networks, which are formed by the same agents, are subsequently conceived as layers, and then a single MLN is generated. The MLN can be formally described as a graph $G = (V, E, J)$, where $V = \{1, 2, ..., x, ..., N\}$ is the set of N agents (i.e. the nodes of the graph), $E = \{1, 2, ..., m, ..., M\}$ is the set of M edges, and $J = \{j_1, j_2, j_3\}$ is the set of the three considered layers. As any edge is located in one and only one layer, mutually exclusive subsets of E can be created so as to separate the edges depending on the layer to which they belong. Formally, we have that $E_j = \{m \in E : \Gamma(m) = j\} \ \forall j \in J$, where Γ is the function that univocally assigns any edge m to its corresponding layer j.

The scope of this MLN is to represent three dimensions of interaction in which the agents are involved. These three dimensions acknowledge socio-economic and biological complex system theories regarding the formation of bottom-up meso-structures [5,20,21,23,25]. In particular, the three aspects that these theories address as fundamental are: *processes*, *structures* and *functions*. With respect to our work, the *processes* are determined by observed agents' co-participations in patents, which identify occurred exchanges of information. The *structures* are determined by the agents' location, which can determine concentration of specific technologies and related know-how, according to theories regarding economic districts [4,9,26]. The *functions* are determined by the agents' technological orientations, which can reveal agents' semantic proximities and convergence towards similar economic processes [11,13,16]. The objective of this MLN is to represent agents in a multi-dimensional space and, more specifically, to infer their potential interaction intensity based on their proximity.

Community Detection of Interactive Economic Agents. In order to identify groups of multi-dimensional interactive agents, a community detection information theory based algorithm, namely Infomap [18,24,28], is implemented on the MLN. The objective is the detection of groups of nodes, i.e. communities of agents, that are likely to exchange information. The community detection algorithm is selected not only for its implementability in MLNs [12], but also for its core and basic functioning. The Infomap algorithm performs the detection of subsets of agents by simulating a spread of a flow throughout the MLN, and minimizing the information needed to describe the circulation of the flow. This allows the identification of meso-structures within which intense information exchanges occur. As information management is an essential organizing principle in the initial formation and in the life-cycle of economic and biological systems [2,6,10,14,15,17,29], the groups detected by Infomap are likely to activate and/or join adaptive bottom-up dynamics.

In order to balance the weight of the three layers, the weights of the connections of each layer are scaled based on the following a ratio. This ratio is computed between the sum of the weights of the connections that belong to the considered layer, and the sum of the connections that belong to a selected layer that is used as reference. For any edge $m \in E$, a weight $\varphi(m)$ that defines the intensity of the connection, is considered. The sum of the weight of the edges of any layer can be computed as $\varphi(E_j) = \sum_{m \in E_j} \varphi(m)$. In addition, with $\varphi^* = \max_{j \in J} \varphi(E_j)$ we refer to the weight of the 'heaviest' layer. Then, in order to normalize the layers' weight, we compute a new edges' weight, namely $\tilde{\varphi}(m)$, as follows

$$\tilde{\varphi}(m) = \varphi(m) \cdot \frac{\varphi^*}{\varphi(E_j)} \tag{1}$$

where $m \in E_j$. As a result, we have that $\tilde{\varphi}(E_{j_1}) = \tilde{\varphi}(E_{j_2}) = \tilde{\varphi}(E_{j_3})$, were $\tilde{\varphi}(E_j)$ is defined specularly to $\varphi(E_j)$. Qualitatively speaking, the sum of the edges' new weight of any layer is equal to the sum of the new edges' weight of any other layer. $\tilde{\varphi}(m)$ is the set of intra-layer edges' weight that is considered when implementing the community detection algorithm[1]. Thanks to this normalisation, the flow that is simulated to circulate throughout the entire MLN structure has the same probability to move to any layer.

The result of this part of analysis is that each agent x is assigned univocally to a community c, as the algorithm is set to search for a hard-partition of G (i.e. non-overlapping subsets of V). In addition, Infomap provides also the information regarding the percentage of simulated flow crossing each agent, namely q.

Topic Modeling of Technological Artifacts. In parallel, a second methodology is developed to investigate the *agent-artifact* space from the artifacts' perspective. In particular, in order to explore their semantic content, an unsupervised learning algorithm is implemented. The unsupervised generative model that is used is the Latent Dirichlet Allocation (LDA) [8]. LDA classifies the collected discrete textual information to a finite number of thematic topics, with the conjecture that they represent combinations of technological subdomains. In this scope, topic per document and words per topic models are established to obtain the most probable thematic groups, or else topics, with Dirichlet multinomial distributions. The assumptions that are made are the following:

- textual information, stored as documents, is a mixture of one or multiple topics simultaneously (as in natural language words may belong to multiple topics), which generate relevant words based on probability distributions
- a topic is a mixture of words from several documents [7,22,30]
- words' order is not considered (exchangeability and bag-of-word assumptions) [1,8,30].

This analysis provides a set of $\theta_k(y)$, each of which represent the probability that the activity y belongs to topic k, using the textual information provided by the activity. For each identified topic, the corresponding θ_k is computed.

3   Analysis of Hierarchical Order in the Distribution of Communities and Topics

The second part of the outlined methodology is made of three stages, which are described in the paragraphs contained in this section.

---

[1] Regarding inter-layer edges, i.e. the edges connecting the layers, these are determined by connections between different state-nodes of a same physical node (i.e. an agent) [12]. In this work, their weight is set equal to the minimum value of $\tilde{\varphi}(E)$, i.e. the set of the intra-layer connections' normalized weights.

The Involvement of Communities in Topics: $w_{c,k}$. In order to assess the involvement of interactive communities of agents in activities' technological sub-domains, we compute the following statistic:

$$w_{c,k} = \sum_{x \in c} q(x) * \sum_{y \in Y_x} \xi(y) * \theta_k(y) \tag{2}$$

where x is an agent belonging to community c, Y_x is the set of activities in which agent x is involved, y is an activity for which $y \in Y_x$, ξ is the fractional counting of the activities, q is the Infomap flow associated with the agents by the analysis of the MLN, and θ_k is the probability that an activity belongs to topic k, as computed by the topic modelling. The fraction counting, i.e. ξ, is a function that computes the reciprocal of the number of agents involved in that activity[2]. This function is used (i) to equally distribute the weight of the activity among all the agents involved and, (ii) to ensure that all the activities have the same weight in the system. In fact, if this was not implemented, the entire weight of an activity would have been repeated for all the agents involved in it. Finally, the values of $w_{c,k}$ are linearly scaled in the interval $[0, 1]$. In this way, the values $\tilde{w}_{c,k}$ are obtained.

The Binary Matrix B_h and the Nestedness Temperature T. The W matrix is determined, with c indicating the rows, k the columns, and $\tilde{w}_{c,k}$ the value of the cells. Then, the presence of topics in communities, as described by matrix W, is computed in a binary way using a threshold. If the presence of a topic in a community, i.e. $\tilde{w}_{c,k}$, is above a threshold h, where $h \in \mathbb{R}$ and $0 \le h \le 1$, then the topic is considered to belong to this community. Communities with all topics below the threshold are discarded, hence not considered to be part of the matrix. This allows to focus on the communities with a minimum strength in terms of involvement in a topic. Based on this step, the binary matrix B_h is generated from the matrix W, depending on the selected value of h. Formally, the value of the elements of the matrix B_h is 1 if the corresponding $\tilde{w}_{c,k} \ge h$, otherwise is 0.

Subsequently, the nestedness temperature T of the matrix B_h is computed, as defined by Atmar and Patterson [3]. Rows and columns of the matrices are sorted in decreasing order by the row-sums and the column-sums, respectively. T which is within the range of $[0, 100]$ degrees, measures the unexpectedness of the non-empty cells that are detected below the anti-diagonal, and the unexpectedness of the empty cells that are detected above the anti-diagonal. These cells are considered as unexpected, as they contribute to increase the disorder in the matrix[3]. The larger the disorder, the higher the nestedness temperature T.

The Statistical Significance of T. In order to assess the statistical significance of the computed nestedness temperature T, homogeneous systems are

---

[2] If the activity y_1 is performed by 4 agents, then $\xi(y_1)$ equals 0.25. If the activity y_2 is performed by 1 agent, then $\xi(y_2)$ equals 1.

[3] A perfectly ordered matrix would have all the non-empty cells above the anti-diagonal, and all the empty cells below the anti-diagonal.

used. Starting from the binary matrix B_h, 1,000 homogeneous matrices B'_h are computed. In order to be homogeneous with respect to B_h, each B'_h presents the following characteristic: the number of communities to which each specific topic belongs is the same as observed in B_h. This means that B'_h are matrices that are randomly generated, with the only constraint that the column-sums are equal to the column-sums of B_h. The matrices B'_h represent homogeneous systems because (i) each topic belongs to the same number of communities as in the original system (B_h), and (ii) the communities to which the topics belong are randomly determined. This stage allows the creation of a sample of matrices to be used to investigate the significance of the nestedness temperature of the original system. The distribution of the values of the nestedness temperature T of the matrices B'_h is used to compute the z-scores of the nestedness temperature T of the matrix B_h. This statistic, namely $z_{T(B_h)}$, is calculated as follows:

$$z_{T(B_h)} = \frac{T(B_h) - \langle T(B'_h) \rangle}{\sigma(T(B'_h))} \tag{3}$$

where $\langle T(B'_h) \rangle$ is the average temperature of B'_h matrices, and $\sigma(T(B'_h))$ is the standard deviation of the temperature of B'_h matrices.

4   The Analysis over the Complex System of Photonics Patents in 2000–2014

Based on a tech-mining approach, a set of 4,926 patents in the field of 'photonics' of the period 2000–2014, are identified. This allows the initial definition of our system, in which the agents are the 1,313 economic institutions (e.g. firms, research institutes and governmental institutions) that were filing at least one the detected patents. As the considered time period is sufficiently large to be cut in smaller time spans, five distinct systems are defined, each of them referring to a different three-year period. The first system refers to the patents, and the corresponding agents, filed in the period 2000–2002, the second system to 2003–2005, the third to 2006–2008, the fourth to 2009–2011, and the fifth to 2012–2014. For each of the five considered systems, a MLN is generated using co-participations in patents to build the first layer (i.e. the one representing *processes* in the *agent-artifact* space), the subregion as geographical level on which to build the second layer (i.e. the one representing *structures*), and the use of common keywords in the filed patents to build the third layer (i.e. the one representing *functions*). The weights of the connections are normalised according to what described in Eq. 1. MLN community detection analyses are implemented (1,000 simulations in any community detection), resulting in the identification of 65, 71, 79, 97 and 78 communities respectively (starting from the system referred to the period 2000–2002, to the system referred to the period 2012–2014). In parallel, the topic modelling analysis, based on all the collected documents, allows the identification of 15 topics.

According to Eq. 2, the computation of the values $\tilde{w}_{c,k}$ is performed for each considered system, so as to obtain the corresponding W matrices. Then, in order

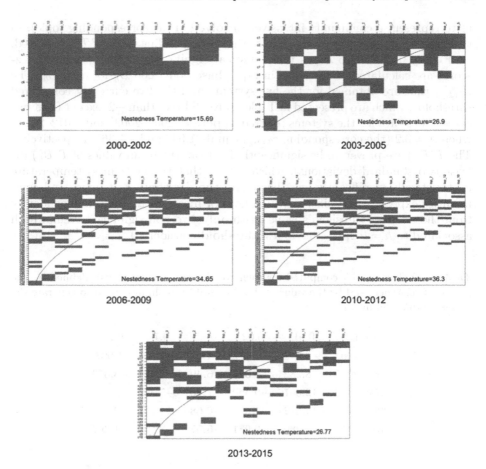

Fig. 1. The five matrices represent the distribution of topics (columns) over communities (rows). The red cells indicate that the topic is developed by the corresponding community. Each matrix presents the result of the community detection in a different time span of the considered system, as indicated below each matrix. The anti-diagonals are represented by black curves. Corresponding levels of detected nestedness temperature are indicated in the bottom-right corner of each matrix. These matrices are all obtained by using a threshold equal to 0.05. This threshold determines the binary allocation of topics to communities, by comparing it to the linearly scaled (in interval $[0, 1]$) weight that measures the involvement of communities in topics. The statistical significance of the obtained nestedness temperatures is independent from which threshold is used out of the five considered. (Color figure online)

to generate the binary matrices B_h, five possible values of the threshold h are considered, namely 0.01, 0.02, 0.05, 0.1 and 0.2. For each of them, nestedness temperature T is calculated. In Fig. 1, the five matrices B_h, with h equal to 0.05, are represented with empty cells in white (so as to represent a community non intensively involved in the corresponding topic), and non-empty cells in red (so as to represent a community intensively involved in the corresponding topic). In the same figure, in each matrix B_h also the corresponding value of T is reported.

Finally, in order to assess the statistical significance of the obtained T, z-scores are calculated as described in Eq. 3, based on 1,000 B_h' for each B_h. The $z_{T(B_h)}$ that are obtained for the five systems and the five currently considered thresholds, which are presented in Table 1, are all lower than -2, except from two cases. These are for the systems referred to period '2000–2002' and '2012–2014', when $h = 0.2$ (the corresponding $z_{T(B_h)}$ equals 1.097 and -1.367, respectively). The $T(B_h)$ are proven to be significantly far from the mean values of $T(B_h')$ (at least two standard deviations), which means that the nestedness temperature is statistically significantly low. Therefore, the distribution of topics and communities expresses a significantly high level of hierarchical order, independently from the adopted threshold's value (with only two non significant results both associated to $h = 0.2$, i.e. the largest threshold considered).

Table 1. Values $z_{T(B_h)}$ computed for each considered system (rows), defined by the period of reference, and by the value of the threshold h (columns) used to generate the corresponding B_h matrix.

Period\h	0.01	0.02	0.05	0.1	0.2
2000–2002	-8.079	-7.172	-4.692	-2.658	1.097
2003–2005	-8.282	-8.326	-6.493	-5.984	-3.650
2006–2008	-12.067	-9.560	-7.761	-5.363	-2.475
2009–2011	-13.512	-11.021	-6.686	-4.121	-2.131
2012–2014	-12.280	-9.981	-6.575	-3.476	-1.367

5   Conclusions

For the developed work, an emerging property of the considered complex system is detected and its statistical significance is confirmed. The analyses reveal that (i) the topics with the minimum diffusion are associated to the communities with the largest set of topics, and (ii) the topics with the maximum diffusion are the only ones to be included in mono-topic communities. Given the obtained results, the methodological approach here outlined to model the *agent-artifact* space is proven to be pertinent to detect an emergent property of it, at least for the considered case study. More specifically, this work provides statistical proofs to support the presence of hierarchical order in the distribution of technological subdomains (i.e. the topics) representing distinct types of artifacts, over interactive groups of agents (i.e. the communities) representing areas of the system hosting intense exchanges of information.

The developed analysis has to be discussed in the context of economic theories, regarding the distribution of different technological topics over different communities belonging to the same *agent-artifact* space. Further analyses will consider the dynamic of the system, as the communities detected in each period do not necessarily have continuity with those detected in the following period. In fact, the presented work addresses the static analysis of five instances of the same system. In addition, the 'unexpected' combinations observed in matrices B_h is inviting to investigate the innovative dynamics. In this perspective, the topological proximities of communities, as observed in the MLN, and the thematic proximities of topics will be considered in a subsequent analysis. Finally, other case studies will be evaluated following the same methodological approach outlined in this work.

References

1. Aldous, D.J.: Exchangeability and related topics. In: Hennequin, P.L. (ed.) École d'Été de Probabilités de Saint-Flour XIII—1983. LNM, vol. 1117, pp. 1–198. Springer, Heidelberg (1985). https://doi.org/10.1007/BFb0099421
2. Arrow, K.J.: Economic welfare and the allocation of resources for invention. In: Readings in Industrial Economics, pp. 219–236. Springer, Berlin (1972). https://doi.org/10.1007/978-1-349-15486-9_13
3. Atmar, W., Patterson, B.D.: The measure of order and disorder in the distribution of species in fragmented habitat. Oecologia **96**(3), 373–382 (1993)
4. Becattini, G.: The Marshallian industrial district as a socio-economic notion. Ind. Dist. Inter-firm Co-op. Italy, 37–51 (1990)
5. Bedau, M.A., Packard, N.H., Rasmussen, S.: Protocells: Bridging Nonliving and Living Matter. MIT Press, Cambridge (2009)
6. Bergstrom, C.T., Lachmann, M.: Shannon information and biological fitness. In: Information Theory Workshop, pp. 50–54. IEEE (2004)
7. Blei, D.M., Lafferty, J.D.: Topic models. Text Min.: Classif. Clust. Appl. **10**(71), 34 (2009)
8. Blei, D.M., Ng, A.Y., Jordan, M.I.: Latent Dirichlet allocation. J. Mach. Learn. Res. **3**(Jan), 993–1022 (2003)
9. Brusco, S.: The emilian model: productive decentralisation and social integration. Camb. J. Econ. **6**(2), 167–184 (1982)
10. Coase, R.H.: The nature of the firm. Economica **4**(16), 386–405 (1937)
11. Cronen, V.E., Pearce, W.B., Harris, L.M.: The logic of the coordinated management of meaning: a rules-based approach to the first course in interpersonal communication. Commun. Educ. **28**(1), 22–38 (1979)
12. De Domenico, M., Lancichinetti, A., Arenas, A., Rosvall, M.: Identifying modular flows on multilayer networks reveals highly overlapping organization in interconnected systems. Phys. Rev. X **5**(1), 011027 (2015)
13. Ding, Y.: Community detection: topological vs. topical. J. Informetr. **5**(4), 498–514 (2011)
14. Donaldson-Matasci, M.C., Bergstrom, C.T., Lachmann, M.: The fitness value of information. Oikos **119**(2), 219–230 (2010)
15. Georgescu-Roegen, N.: The Entropy Law and the Economic Process. Harvard University Press, Cambridge (1971)
16. Hacklin, F.: Management of Convergence in Innovation: Strategies and Capabilities for Value Creation Beyond Blurring Industry Boundaries. Springer, Berlin (2007). https://doi.org/10.1007/978-3-7908-1990-8

17. Hicks, J.R.: Capital and Time: A Neo-Austrian Theory. Clarendon Press, Oxford (1973)
18. Huffman, D.A.: A method for the construction of minimum-redundancy codes. Proc. IRE **40**(9), 1098–1101 (1952)
19. Lane, D., Maxfield, R., et al.: Foresight, Complexity, and Strategy. Santa Fe Institute, New Mexico (1997)
20. Lane, D.A.: Complexity and innovation dynamics. Handb. Econ. Complex. Technol. Change **63** (2011). Edward Elgar Publishing, Cheltenham
21. Lane, D.A., Maxfield, R.R.: Ontological uncertainty and innovation. J. Evol. Econ. **15**(1), 3–50 (2005)
22. Papadimitriou, C.H., Raghavan, P., Tamaki, H., Vempala, S.: Latent semantic indexing: a probabilistic analysis. J. Comput. Syst. Sci. **61**(2), 217–235 (2000)
23. Righi, R.: A methodological approach to investigate interactive dynamics in innovative socio-economic complex systems. Stat. Appl. - Italian J. Appl. Stat. **30**(1), 113–142 (2018)
24. Rosvall, M., Bergstrom, C.T.: Maps of random walks on complex networks reveal community structure. Proc. Natl. Acad. Sci. **105**(4), 1118–1123 (2008)
25. Samoili, S., Righi, R., Lopez-Cobo, M., Cardona, M., De Prato, G.: Unveiling latent relations in the photonics techno-economic complex system. In: Cagnoni, S., Mordonini, M., Pecori, R., Roli, A., Villani, M. (eds.) WIVACE 2018. CCIS, vol. 900, pp. 72–90. Springer, Cham (2019). https://doi.org/10.1007/978-3-030-21733-4_6
26. Saxenian, A.: Regional networks: industrial adaptation in silicon valley and route 128 (1994)
27. Selva, N., Fortuna, M.A.: The nested structure of a scavenger community. Proc. R. Soc. B: Biol. Sci. **274**(1613), 1101–1108 (2007)
28. Shannon, C.E.: A mathematical theory of communication, part I, part II. Bell Syst. Tech. J. **27**, 623–656 (1948)
29. Smith, J.M.: The concept of information in biology. Philos. Sci. **67**(2), 177–194 (2000)
30. Steyvers, M., Griffiths, T.: Probabilistic topic models. Handb. Latent Semant. Anal. **427**(7), 424–440 (2007)

Towards Programmable Chemistries

Dandolo Flumini[1]([⊠]), Mathias S. Weyland[1], Johannes J. Schneider[1],
Harold Fellermann[2], and Rudolf M. Füchslin[1]

[1] School of Engineering, Zurich University of Applied Sciences,
Technikumstrasse 9, 8400 Winterthur, Switzerland
{flum,weyl,scnj,furu}@zhaw.ch
[2] School of Computing, Newcastle University,
Newcastle-upon-Tyne NE4 5TG, UK
harold.fellermann@ncl.ac.uk

Abstract. We provide a practical construction to map (slightly modified) *GOTO*-programs to chemical reaction systems. While the embedding reveals that a certain small fragment of the chemtainer calculus is already Turing complete, the main goal of our ongoing research is to exploit the fact that we can translate arbitrary control-flow into real chemical systems. We outline the basis of how to automatically derive a physical setup from a procedural description of chemical reaction cascades. We are currently extending our system in order to include basic chemical reactions that shall be guided by the control-flow in the future.

Keywords: Programmable chemistry · Compartmentalization ·
Biochemical engineering · Theoretical computer science

1 Introduction

In order to "program" chemical reaction systems, we provide a construction to map procedural control-flow to chemical reaction systems.

The computational framework that we use to represent arbitrary control flow is (slightly modified) GOTO programs. In Sect. 3 we give a short introduction to the GOTO formalism. The results presented in this section are standard. The related result presented in Lemma 1 (Sect. 4) is also considered to be known; however, no corresponding reference was found.

The system to represent chemical reaction systems is the chemtainer calculus. In Sect. 2, we give a short introduction to the relevant notions of the formalism. For a more detailed account, the reader is referred to [16].

Section 4 is the main contribution of the present work. We discuss the actual embedding of arbitrary GOTO programs into the chemtainer calculus. We also

Funded by Horizon 2020 Framework Programme of the European Union (project 824060).

The original version of this chapter was previously published non-open access. A Correction to this chapter is available at https://doi.org/10.1007/978-3-030-45016-8_18

F. Cicirelli et al. (Eds.): WIVACE 2019, CCIS 1200, pp. 145–157, 2020.
https://doi.org/10.1007/978-3-030-45016-8_15

discuss a variation of the construction that solves some issues that render the original embedding unsuitable for practical use in a "programmable chemistry" setting. All of the work presented is original research.

In Sect. 5 we discuss our results, ongoing research, and future directions.

1.1 Related Work

Chemical reaction systems are formally described by chemical reaction networks (CRNs) [1,2]. They can be used to facilitate the analysis of artificial chemistries [3] and real chemistries. To name a few applications, CRNs have been used to predict reaction paths [4], to model spontaneous emergence of self-replication [5], to synthesize optimal CRNs from prescribed dynamics [6], and to design asynchronous logic circuits [7].

In the European Commission-funded project ACDC, we are developing a programmable artificial cell with distributed cores. An important feature of the systems studied in the context of ACDC is compartmentalization. CRNs alone are not suitable to model compartmentalization. However, formalisms with the ability to express compartmentalization have been developped [8–16]. Our systems can be described particularly well with the chemtainer calculus [16], one of the aforementioned formalisms. Thus, the chemtainer calculus is chosen for emulating computations with chemical reaction systems in the present work.

2 The Chemtainer Calculus

As discussed in the previous section, the chemtainer calculus is a formal calculus capable of describing compartmentalized reaction systems [16]. In this section, the subset of chemtainer calculus necessary for the emulation of computations with chemical reaction systems is introduced. It consists of the following objects:

- **Molecules:** Objects that can undergo reactions as specified in a CRN. Capital letters $(A, B, C, ...a)$ are used to denote molecules.
- **Chemtainers:** Compartments that contain objects (including other chemtainers). The symbols ⦇ and ⦈ are used to indicate objects enclosed in chemtainers.
- **Address tags:** Tags that are in solution or attached to a chemtainer. Lower case greek letters $(\tau, \sigma, ...)$ are used to denote tags.

The notion of space is implemented with discrete locations $(x, y, m_i, ...)$ at which objects reside. A number of instructions are used to alter the system state. They are introduced below by examples:

- **feed**$(x, A, 3)$: A chemtainer containing 3 instances of molecule A is fed into location x. Starting from an empty state, this yields:

$$\emptyset \rightarrow x : ⦇\, 3A \,⦈$$

- **feed_tag**$(x, \sigma, 1)$: One tag with address σ is fed into location x. Again presuming an empty initial state, the instruction yields:

$$\emptyset \to x : 1\sigma$$

- **tag**(x): Decorate chemtainer in location x with the tags surrounding the chemtainer:

$$x : 1\sigma + (\!(\, 3A \,)\!) \to x : 1\sigma(\!(\, 3A \,)\!)$$

- **move**(σ, x, y): Move any tag σ (including potentially attached chemtainers) from location x to location y:

$$x : 1\sigma(\!(\, 3A \,)\!) \to y : 1\sigma(\!(\, 3A \,)\!)$$

- **fuse**(x): Fuse chemtainers in location x:

$$x : (\!(\, 2A \,)\!) + (\!(\, 2B \,)\!) \to x : (\!(\, 2A + 2B \,)\!)$$

- **flush**(x): Remove any objects from location x:

$$x : (\!(\, 2A + 2B \,)\!) \to x : \emptyset$$

- **burst**(x): Burst chemtainers in location x, releasing any contained molecules and leaving behind empty chemtainers:

$$x : (\!(\, 2A + 2B \,)\!) \to 2A + 2B + (\!(\, \,)\!)$$

Chemtainer programs are a sequence of such instructions that alter the system state.

3   GOTO-Programs

We work with a slight variation of the standard syntax of *GOTO* programs as presented in [17]. Our syntactic building blocks are as follows:

- Countably many **variables** $x_0, x_1, x_2, \ldots$,
- **literals** $0, 1, 2, \ldots$ for nonnegative integers,
- **markers** $M_1, M_2, M_3, \ldots$,
- **separator** symbols $=, :$,
- **operator** and **relation**symbols $+, -, >$,
- and **keywords** GOTO, IF, THEN, HALT.

Instructions of GOTO-programs take one of the following forms:

- **Assignments:** $x_i := x_i \pm c$ where $i \in \mathbb{N}$, c is a literal (for a nonnegative integer), and $\pm$ stands for either $+$ or $-$.
- **Jumps:** GOTO M_k where $k \in \mathbb{N}$
- **Conditional Jumps:** IF $x_i > 0$ THEN GOTO M_k where $i, k \in \mathbb{N}$
- **Halt instruction:** HALT.

A GOTO-program is a finite sequence of instructions, each of which is given with a unique label of the form M_i where $i \in \mathbb{N}$. In order to enhance readability, we will generally write up GOTO-programs in vertical order

$$M_1 : I_1$$

$$\vdots$$

$$M_k : I_k.$$

In Sect. 4, we observe how every computation of a *GOTO*-program can be "emulated" by the state-transitions of a suitably constructed chemical reaction system. As a result, we obtain that a suitable chemical reaction system can emulate every computation (in the sense of Turing completeness).

To match *GOTO*-program computations with state-transitions of a chemical reaction system, we introduce an operational semantics for the *GOTO* language that captures the idea of a global state being mutated while instructions are executed sequentially.

The state of a *GOTO*-program-computation is completely determined by a marker (stating the "current" instruction) and the values held by relevant variables (i.e. all variables occurring in the program at hand). Thus, for a given *GOTO*-program P with variables $x_0, \ldots, x_n$ and markers $M_1, \ldots, M_k$, the state of a computation can be modelled as a tuple $(X, \boldsymbol{y})$ where $X \in \{M_1, \ldots, M_k, \bot\}$ indicates the "current" instruction and $\boldsymbol{y} \in \mathbb{N}^{n+1}$ holds the values stored in the variables $x_0, \ldots, x_n$. Cases where $X = \bot$ indicate that the computation has halted. The (deterministic) operational semantics is given by the following transition relation:

Let P be any *GOTO*-program with variables among $x_0, \ldots, x_n$, let $y_0, y_1, \ldots$ range over natural numbers, and let l_c stand for the literal associated with a natural number c.

- If $M_i : x_r := x_r \pm l_c$ is part of P, and if the following line is labelled with marker M_k, then

$$(M_i, y_0, \ldots, y_r, \ldots, y_n) \xrightarrow{\;P\;} \begin{cases} (M_k, y_1, \ldots, 0, \ldots, y_n) & \text{if } y_r \pm c \leq 0 \\ (M_k, y_1, \ldots, y_r \pm c, \ldots, y_n) & \text{otherwise.} \end{cases}$$

- If $M_i : x_r := x_r \pm l_c$ is the last line of P, then

$$(M_i, y_0, \ldots, y_r, \ldots, y_n) \xrightarrow{\;P\;} \begin{cases} (\bot, y_1, \ldots, 0, \ldots, y_n) & \text{if } y_r \pm c \leq 0 \\ (\bot, y_1, \ldots, y_r \pm c, \ldots, y_n) & \text{otherwise.} \end{cases}$$

- If $M_i : \text{GOTO } M_k$ is a line of P, then

$$(M_i, y_0, \ldots, y_n) \xrightarrow{\;P\;} \begin{cases} (M_k, y_0, \ldots, y_n) & \text{if } M_k \text{ is a label in } P \\ (\bot, y_0, \ldots, y_n) & \text{otherwise} \end{cases}$$

– If M_i : IF $x_j > 0$ THEN GOTO M_k is a line in P, and if $y_j > 0$, then

$$(M_i, y_0, \ldots, y_n) \xrightarrow{\ P\ } \begin{cases} (M_k, y_0, \ldots, y_n) & \text{if } M_k \text{ is a label in } P \\ (\perp, y_0, \ldots, y_n) & \text{otherwise.} \end{cases}$$

– If M_i : IF $x_j > 0$ THEN GOTO M_k is a line in P, and if $y_j = 0$ and the following line is labelled with marker M_k, then

$$(M_i, y_0, \ldots, y_n) \xrightarrow{\ P\ } (M_k, y_0, \ldots, y_n)$$

– If M_i : IF $x_j > 0$ THEN GOTO M_k is the last line in P, and if $y_j = 0$, then

$$(M_i, y_0, \ldots, y_n) \xrightarrow{\ P\ } (\perp, y_0, \ldots, y_n).$$

– If M_k : HALT is a line in P, then

$$(M_k, y_0, \ldots, y_n) \xrightarrow{\ P\ } (\perp, y_0, \ldots, y_n)$$

– No other cases are considered.

Further, we write $x \xrightarrow{P, 1} y$ if $x \xrightarrow{\ P\ } y$, and $x \xrightarrow{P, n+1} y$ if there is a state z such that $x \xrightarrow{P, n} z$ and $z \xrightarrow{\ P\ } y$. Since labels in G-programs are unique, the resulting transition system is deterministic in the sense that $x \xrightarrow{\ P\ } y \wedge x \xrightarrow{\ P\ } y'$ implies $y = y'$ for all states x, y and y'. Therefore it is meaningful to write $x^{P,n}$ for the unique state of P that satisfies $x \xrightarrow{P, n} x^{P,n}$.

Based on the given transition relation we can introduce the usual denotational semantics for GOTO-programs; for every GOTO-program P and every $k \in \mathbb{N}$ the (partial) function $[P, k] : \mathbb{N}^k \to \mathbb{N}$ is given from:

$$[P, k](y_1, \ldots, y_k) = y \Leftrightarrow \exists n, z ((m, (0, y_1, \ldots, y_k, \mathbf{0}))^{P,n} = (\perp, (y, z)) \qquad (1)$$

where m denotes the marker of the first line in P and $\mathbf{0}$ represents a sequence of zeros, so that all variables in P are initialized properly. The equivalence stated in (1) means that we evaluate a *GOTO* program P as a k-ary $[P, k]$ function as follows:

– Initialize the variables $x_1, \ldots, x_k$ with the input values (additional variables of P are initialized with 0).
– Execute the program P starting with the first instruction and according to the state transitions given above.
– If the execution halts, read the variable x_0 to obtain the output of the function $[P, k]$ for the given input vector.

It is well known that *GOTO* is Turing complete with respect to this semantics [17].

4   Emulating Computations with Chemical Reaction Systems

In this section, we rely on the notions of artificial cellular matrices, the chemtainer calculus as introduced in Sect. 2, and chemtainer programs. We refer the reader to [16] for further details.

We demonstrate how given any $GOTO$-program P, we can construct an artificial cellular matrix M together with a chemtainer program to simulate P. In a first step, we will show how to match any state x of a $GOTO$-program to a global state $\ulcorner x \urcorner$ of the chemtainer calculus, and then we will describe how to translate the $GOTO$ program P into a corresponding chemtainer program $\langle\!\langle P \rangle\!\rangle$, so that all state transitions of P are simulated in M (Proposition 1).

4.1   Matching States of $GOTO$-Program-Computations with Global States of the Chemtainer Calculus

For a given $GOTO$-program P with variables $x_0, \ldots, x_n$ and markers $M_1, \ldots, M_k$, we identify states $(M_i, \boldsymbol{y})$ of the computations of P with global states $\ulcorner (M_i, \boldsymbol{y}) \urcorner$ of the chemtainer calculus as follows: We use tags $\tau_0, \ldots, \tau_n$, a special "control-flow" tag σ, and locations $\tilde{m}_0, m_1, \tilde{m}_1 \ldots, m_k, \tilde{m}_k$ as well as a special location $halt$ to stipulate

$$\ulcorner (M_i, \boldsymbol{y}) \urcorner =$$
$$\tilde{m}_0 : 0 \circ \tilde{m}_1 : 0 \circ m_1 : 0 \circ \cdots \circ m_i : \tau_0^{y_0} \ldots \tau_n^{y_n} \llparenthesis\, 0\, \rrparenthesis \circ \cdots \circ m_k : 0 \circ halt : 0$$

and

$$\ulcorner (\bot, \boldsymbol{y}) \urcorner = \tilde{m}_0 : 0 \circ \tilde{m}_1 : 0 \circ m_1 : 0 \circ \cdots \circ m_k : 0 \circ halt : \tau_0^{y_0} \ldots \tau_n^{y_n} \llparenthesis\, 0\, \rrparenthesis$$

where $\tau_i^{y_i}$ stands for the y_i fold repetition of τ_i. An illustration of the correspondence is shown in Fig. 1.

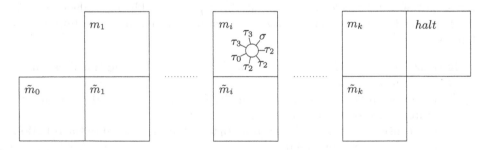

Fig. 1. An illustration of the state $(M_i, 1, 3, 2, 0, \ldots, 0)$ of a $GOTO$-program-computation interpreted as a global state of the chemtainer calculus.

4.2   The Construction of the Chemtainer Program

The mapping of states defined in Sect. 4.1 now enables us to specify a construction enabling us to translate any $GOTO$ program P to a chemtainer program $\langle\!\langle P \rangle\!\rangle$ that emulates the computation of P. Our general strategy is first to associate lines (i.e., tagged instructions) $M_j : I_j$ of the $GOTO$-language to simple chemtainer programs $\langle M_j : I_j \rangle$, and then to show how the mapping can be extended to translate complete $GOTO$ programs consisting of several lines of code.

The basic chemtainer programs $\langle M_j : I_j \rangle$ are specified by case analysis as follows:

$$\langle M_j : x_i := x_i + l_c \rangle = \textbf{feed\_tag}(m_j, \tau_i, c); \textbf{tag}(m_j)$$
$$\langle M_j : x_i := x_i - l_c \rangle = \textbf{feed\_tag}(m_j, \bar{\tau}_i, c)$$
$$\langle M_j : \textsf{GOTO}\ M_i \rangle = \textbf{move}(\sigma, m_j, \tilde{m}_{i-1})$$
$$\langle M_j : \textsf{IF}\ x_r > 0\ \textsf{THEN GOTO}\ M_i \rangle = \textbf{move}(\tau_r, m_j, \tilde{m}_{i-1})$$
$$\langle M_i : \textsf{HALT} \rangle = \textbf{move}(\sigma, i, halt).$$

Next, we translate $GOTO$ programs that are composed of several instructions. In favor of a more concise description, we will here and henceforth assume (without loss of generality) that $GOTO$ program-lines are marked in order $M1, M2, M3, \ldots$, and that jump instructions may only lead to markers that are present in the program at hand. We thus assume, that the given program P is of the form

$$M_1 : I_1$$
$$\vdots$$
$$M_k : I_k,$$

and we stipulate $\langle\!\langle P \rangle\!\rangle$ for the following chemtainer program:

$\langle M_1 : I_1 \rangle;$
$\langle M_2 : I_2 \rangle;$
$\vdots$
$\langle M_k : I_k \rangle;$
$\textbf{move}(\sigma, m_1, \tilde{m}_1); \ldots; \textbf{move}(\sigma, m_k, \tilde{m}_k);$
$\textbf{flush}(m_1); \ldots; \textbf{flush}(m_k)$
$\textbf{move}(\sigma, \tilde{m}_0, m_1); \textbf{move}(\sigma, \tilde{m}_1, m_2); \ldots; \textbf{move}(\sigma, \tilde{m}_k, halt);$

Now, given a global state S of the chemtainer calculus, we write $S^{P,n}$ for the global state (in order to obtain determinism, we here need to restrict the original rule number 56 of the chemtainer calculus (as introduced in [16]) to only be admissible if the respective location is empty.) that results from S when the program $\langle\!\langle P \rangle\!\rangle$ is applied exactly n times. In the next lemma, we state that the correspondence declared in Sect. 4.1 is a simulation relation.

Proposition 1. *Let P be any GOTO-program of the form*

$$M_1 : I_1$$

$$\vdots$$

$$M_k : HALT$$

such that all jump instructions in P refer to a marker $M_1, \ldots, M_k$. If x is a state in a computation of P, then for all $n \in \mathbb{N}$,

$$\ulcorner x^{P,n} \urcorner = \ulcorner x \urcorner^{\langle\!\langle P \rangle\!\rangle, n}.$$

Proof. Let P and x be as stated in the claim. Applying induction on n, we only need to prove that $\ulcorner x^{P,1} \urcorner = \ulcorner x \urcorner^{P,1}$. Let x be

$$(M_i, \boldsymbol{y})$$

where $\boldsymbol{y} = y_0, \ldots, y_n$. The proof proceeds by case distinction on the instruction I_i. We can assume that I_i is not the last instruction in P if I_i is not the $HALT$ instruction.

- If I_i is $x_j := x_j + l_c$, then $x^{P,1}$ is $(M_{i+1}, y_0, \ldots, y_j + c, \ldots, y_n)$ and thus

$$\ulcorner x^{P,1} \urcorner = \tilde{m}_0 : 0 \circ m_1 : 0 \circ \tilde{m}_1 : 0 \circ \ldots \circ m_{i+1} : \tau_0^{y_0} \ldots \tau_j^{y_j+c}$$
$$\ldots \tau_n^{y_n} (\!(0)\!) \circ \cdots \circ m_k : 0 \circ halt : 0.$$

  When running $\langle\!\langle P \rangle\!\rangle$ with initial state

$$\ulcorner x \urcorner = \tilde{m}_0 : 0 \circ m_1 : 0 \circ \tilde{m}_1 : 0 \circ \ldots$$
$$\circ \, m_i : \tau_0^{y_0} \ldots \tau_n^{y_n} (\!(0)\!) \circ \cdots \circ m_k : 0 \circ halt : 0$$

  the right number of tags are attached to the chemtainer in the "first part" of the program, and the chemtainer is relocated to m_{i+1} in two steps resulting in the same global state

$$\ulcorner x \urcorner^{\langle\!\langle P \rangle\!\rangle, 1} = \tilde{m}_0 : 0 \circ m_1 : 0 \circ \tilde{m}_1 : 0 \circ \cdots \circ m_{i+1} : \tau_0^{y_0} \ldots \tau_j^{y_j+c}$$
$$\ldots \tau_n^{y_n} (\!(0)\!) \circ \cdots \circ m_k : 0 \circ halt : 0.$$

- The case where I_i is $x_j := x_j - l_c$ works essentially like the previous case, with the difference that no tagging instruction is introduced and the released tags bind to the complementary tags (that are already attached to the chemtainer).
- If I_i is IF $x_j > 0$ THEN GOTO M_r and $y_j = 0$, then the state $x^{P,1}$ is $(M_{i+1}, \boldsymbol{y})$ and thus $\ulcorner x^{P,1} \urcorner$ is

$$\tilde{m}_0 : 0 \circ m_1 : 0 \circ \tilde{m}_1 : 0 \circ \cdots \circ m_{i+1} : \tau_0^{y_0} \ldots \tau_n^{y_n} (\!(0)\!) \circ \cdots \circ m_k : 0 \circ halt : 0$$

On the other hand, if we run the chemtainer program $\langle\!\langle P \rangle\!\rangle$ with starting state

$$\ulcorner x \urcorner = \tilde{m}_0 : 0 \circ m_1 : 0 \circ \tilde{m}_1 : 0 \circ \cdots \circ m_i : \tau_0^{y_0} \ldots \tau_n^{y_n} (\!(\, 0 \,)\!) \circ \cdots \circ m_k : 0 \circ halt : 0,$$

we note that since $y_j = 0$ there is no τ_j on the surface of the chemtainer, thus no transition in the first "half" of $\langle\!\langle P \rangle\!\rangle$ is effective at all. Thus, the only instructions that have an impact on the global state $\ulcorner x \urcorner$ are $\mathbf{move}(\sigma, m_i, \tilde{m}_i)$ and $\mathbf{move}(\sigma, \tilde{m}_i, m_{i+1})$, resulting in the global state $\ulcorner x \urcorner^{P,1} = \ulcorner x^{P,1} \urcorner$.

- If I_i is IF $x_j > 0$ THEN GOTO M_r and $y_j > 0$, then the state $x^{P,1}$ becomes $(M_r, \boldsymbol{y})$ (we assume that the marker M_r exists in P.), and thus

$$\ulcorner x^{P,1} \urcorner =$$
$$\tilde{m}_0 : 0 \circ m_1 : 0 \circ \tilde{m}_1 : 0 \circ \cdots \circ m_r : \tau_0^{y_0} \ldots \tau_n^{y_n} (\!(\, 0 \,)\!) \circ \cdots \circ m_k : 0 \circ halt : 0.$$

Accordingly, if we run the chemtainer-program with initial state $\ulcorner x \urcorner$, the instructions of $\langle\!\langle P \rangle\!\rangle$ that actually alter the global state are $\langle M_i : I_i \rangle$ i.e. $\mathbf{move}(\tau_j, m_i, \tilde{m}_{r-1})$ and $\mathbf{move}(\sigma, \tilde{m}_{r-1}, m_r)$, thus we obtain

$$\ulcorner x \urcorner^{\langle\!\langle P \rangle\!\rangle, 1} =$$
$$\tilde{m}_0 : 0 \circ m_1 : 0 \circ \tilde{m}_1 : 0 \circ \cdots \circ m_r : \tau_0^{y_0} \ldots \tau_n^{y_n} (\!(\, 0 \,)\!) \circ \cdots \circ m_k : 0 \circ halt : 0$$

as desired.
- Nonconditional jump instructions are handled exactly like conditional jump instructions where the condition is satisfied.
- If I_i is $HALT$, then $x^{P,1}$ is $(\bot, \boldsymbol{y})$ and thus $\ulcorner x^{P,1} \urcorner$ is

$$\tilde{m}_0 : 0 \circ \tilde{m}_1 : 0 \circ m_1 : 0 \circ \cdots \circ m_k : 0 \circ halt : \tau_0^{y_0} \ldots \tau_n^{y_n} (\!(\, 0 \,)\!)$$

Since the only relevant transition in $\langle\!\langle P \rangle\!\rangle$ when applied to

$$\ulcorner x \urcorner = \tilde{m}_0 : 0 \circ m_1 : 0 \circ \tilde{m}_1 : 0 \circ \ldots$$
$$\circ m_i : \tau_0^{y_0} \ldots \tau_n^{y_n} (\!(\, 0 \,)\!) \circ \cdots \circ m_k : 0 \circ halt : 0$$

is $\langle M_i : Halt \rangle = \mathbf{move}(\sigma, \mathbf{i}, \mathbf{halt})$ the states $\ulcorner x^{P,1} \urcorner$ and $\ulcorner x \urcorner^{\langle\!\langle P \rangle\!\rangle, 1}$ coincide.

$\square$

As a corollary we obtain that any (Turing) computable function can be evaluated by a suitable artificial cellular matrix together with an expression of the chemtainer calculus.

Corollary 1. *Given any recursive funtion $f : \mathbb{N}^k \to \mathbb{N}$, then an artificial cellular matrix, a natural number n and a chemtainer program P exist such that*

$$(\tilde{m}_0 : \tau_1^{y_1} \ldots \tau_n^{y_n} (\!(\, 0 \,)\!) \circ m_1 : 0 \circ \cdots \circ m_k : 0 \circ halt : 0)^{P,n}$$
$$= \cdots \circ halt : \tau_0^{f(y_1, \ldots, y_n)} \ldots (\!(\, 0 \,)\!)$$

holds whenever $f(y_1, \ldots, y_n)$ is defined.

Proof. This follows from Proposition 1 and Eq. 1. $\square$

4.3 Practical Considerations

While theoretically sound, we identified two main issues of our construction that make it unsuitable for practical use in a "programmable chemistry" setting, both of which have to do with how we encode natural numbers in quantities of molecules:

- If a large number of variables occur in a program, then there might not be a distinct (suitable) molecule for each variable to encode. We call this the "finiteness of molecules problem".
- It is generally infeasible to exactly count numbers of molecules (which means that we cannot effectively read or write variables). We call this the "counting problem".

We can solve the finiteness problem by pointing out that there is a definite natural number N_0 such that every $GOTO$-computation can be realized by a $GOTO$-program with no more than N_0 many variables. This is equivalent to the statement of the next lemma.

Lemma 1. *A natural number N_0 exists, such that for every GOTO program P there is a GOTO program P' with no more than N_0 many variables and $[P,1] = [P',1]$.*

Proof. Since the $GOTO$ language is Turing complete, a $GOTO$ program I exists, such that for a suitable encoding $\#$ of $GOTO$ programs the equation

$$\lambda x.[I,2](\#A, x) = [A,1]$$

holds for every $GOTO$ program A. Thus, given any $GOTO$ program P a suitable $GOTO$ program P' that satisfies the claim is given from

$$M_a : x_2 := x_1 + l_0;$$
$$M_b : x_1 := l_{\#P} + l_0;$$
$$I$$

where the markers M_a and M_b do not occur in I.   □

In order to solve the counting problem, we have to modify our construction slightly. Since exactly counting the numbers of molecules is not feasible, it is not suitable to represent integer values by exactly matching numbers of specific tags on the surface of a chemtainer. It is, however, simple to measure concentrations and thus to decide whether the concentration of a molecule is "high" or "low" respectively. If not integer values, this enables us to code boolean values effectively. The main idea is as follows:

G_{bool}-programs are obtained by restricting constant and variable values in $GOTO$-programs to 0 or 1 respectively. In contrast to our first embedding, if a variable x_i holds the value 1, this is not translated in the sense that there is exactly one tag τ_i on the surface of some vesicle, but rather that there are many i.e., that the vesicles surface is "covered" with corresponding tags. Accordingly, states of G_{bool}-program computations are identified with global states of the chemtainer calculus similar as in Sect. 4 but the $\tau_k^{y_k}$'s stand for a very short string of τ_k if $y_k = 0$ and a very long string of τ_k otherwise. Similar to the situation shown in Fig. 1, the state $(M_i, 0, 1, 0, 1, 1, 0, 0, 0, 0)$ is represented by a chemtainer in location m_i with its surface populated by many σ, τ_1, τ_3 and τ_4 tags and none or very few further tags.

The construction of simple chemtainer programs to emulate labeled instructions of G_{bool}-programs then essentially works as with $GOTO$-programs and is given from

$$\langle M_j : x_i := x_i + c \rangle = \begin{cases} \mathbf{feed\_tag}(m_j, \tau_i, \infty); \mathbf{tag}(m_i) & \text{if } c = 1 \\ \epsilon & \text{otherwise} \end{cases}$$

$$\langle M_j : x_i := x_i - c \rangle = \begin{cases} \mathbf{feed\_tag}(m_j, \bar{\tau}_i, \infty); \mathbf{tag} & \text{if } c = 1 \\ \epsilon & \text{otherwise} \end{cases}$$

$$\langle M_j : \mathsf{GOTO}\ M_i \rangle = \mathbf{move}(\sigma, m_j, \tilde{m}_{i-1})$$

$$\langle M_j : \mathsf{IF}\ x_r > 0\ \mathsf{THEN}\ \mathsf{GOTO}\ M_i \rangle = \mathbf{move}(\tau_r, m_j, \tilde{m}_{i-1})$$

$$\langle M_i : \mathsf{HALT} \rangle = \mathbf{move}(\sigma, i, halt)$$

where $\mathbf{feed\_tag}(m_j, \tau_i, \infty)$ means that the location m_j is flooded with a non-specific but abundant number of τ_i tags. The embedding of a G_{bool} program into the chemtainer calculus remains exactly as in the case of $GOTO$ programs.

5 From a Practical Embedding Towards a Higher Level Programming Language for Chemical Reaction Control

Thus far, we have shown how to map modified GOTO-programs to chemtainer systems and how those systems can simulate the computation of programs. While these embeddings reveal that the chemtainer calculus is indeed Turing complete, this does not come to a great surprise. However, our constructions are explicit; they constitute an algorithm that translates given programs to actual chemtainer systems that can be executed chemically. In that sense, we have outlined the construction of a very simple chemical compiler to capture the control-flow of a simplified programming language in a setup of artificial cellular matrices. Our current focus is now on adding proper chemical operations in the sense of a library to our framework and to continue improving our system to denote intended chemical reactions and products in a more declarative manner. In terms of semantics, we are working on a probabilistic interpretation to capture the nature of chemical reaction systems more accurately.

The embedding we have shown in this work is far away from what can be done in a laboratory. Nevertheless, we claim that our work has some practical implications. From a mathematical perspective, the presented embedding is a solid starting point for further developments. Solid because Turing completeness allows referring to a large body of well- established results. Adding further functions will not change the property of Turing completeness, but only facilitate the implementation (additional functions may be chosen with particular attention to chemical practicability). We aim at a bi-directional way of inspiration. Mathematical consideration may suggest specific functions to be of high usability (e.g., because they facilitate compilation). It is then a question for chemistry whether these functions can be implemented. Going in the other direction, biology provides us with sophisticated mechanisms, e.g. for the synthesis of branched oligosaccharides [18]. Given a mathematical framework, one may ask how to translate such evolved functions into a formal framework and to what extent they offer general tools.

One may even go a step further. In this work, we emphasize Turing completeness. Comparing our embedding to what one finds in biology may shed light on the role of Turing completeness. We don't assume biological systems to exhibit specific mathematical properties; it is, however, of interest to analyze in what respect biological process control differs from the ideal one has constructed in computer science.

Finally, we highlight the difference between procedural and declarative languages. The presented embedding follows the procedural paradigm. However, chemical kinetics is, by its very nature a prime example for a declarative language with a semantics that can be simulated by the Gillespie algorithm [19, 20]. We claim that further progress towards the understanding of biological processes and chemical process control has to include a shift from the procedural to the declarative point of view in order to take account of the fundamental nature of chemistry.

References

1. Feinberg, M.: Some recent results in chemical reaction network theory. In: Patterns and Dynamics in Reactive Media. Springer, New York (1991). https://doi.org/10.1007/978-1-4612-3206-3_4
2. Banzhaf, W., Yamamoto, L.: Artificial Chemistries. MIT Press, Cambridge (2015)
3. Dittrich, P., Ziegler, J., Banzhaf, W.: Artificial chemistries - a review. Artif. Life 7(3), 225–275 (2001)
4. Kim, Y., Kim, J.W., Kim, Z., Kim, W.Y.: Efficient prediction of reaction paths through molecular graph and reaction network analysis. Chem. Sci. 9(4), 825–835 (2018)
5. Liu, Y., Sumpter, D.J.T.: Mathematical modeling reveals spontaneous emergence of self-replication in chemical reaction systems. J. Biol. Chem. 293(49), 18854–18863 (2018)
6. Cardelli, L., et al.: Syntax-guided optimal synthesis for chemical reaction networks. In: Majumdar, R., Kunčak, V. (eds.) CAV 2017. LNCS, vol. 10427, pp. 375–395. Springer, Cham (2017). https://doi.org/10.1007/978-3-319-63390-9_20
7. Cardelli, L., Kwiatkowska, M., Whitby, M.: Chemical reaction network designs for asynchronous logic circuits. Nat. Comput. 17(1), 109–130 (2017). https://doi.org/10.1007/s11047-017-9665-7

8. Nardin, C., Widmer, J., Winterhalter, M., Meier, W.: Amphiphilic block copolymer nanocontainers as bioreactors. Eur. Phys. J. E **4**(4), 403–410 (2001)
9. Noireaux, V., Libchaber, A.: A vesicle bioreactor as a step toward an artificial cell assembly. Proc. Natl. Acad. Sci. **101**(51), 17669–17674 (2004)
10. Roodbeen, R., Van Hest, J.C.: Synthetic cells and organelles: compartmentalization strategies. BioEssays **31**(12), 1299–1308 (2009)
11. Baxani, D.K., Morgan, A.J., Jamieson, W.D., Allender, C.J., Barrow, D.A., Castell, O.K.: Bilayer networks within a hydrogel shell: a robust chassis for artificial cells and a platform for membrane studies. Angewandte Chemie Int. Ed. **55**(46), 14240–14245 (2016)
12. Li, J., Barrow, D.A.: A new droplet-forming fluidic junction for the generation of highly compartmentalised capsules. Lab Chip **17**(16), 2873–2881 (2017)
13. Păun, G.: Computing with membranes. J. Comput. Syst. Sci. **61**(1), 108–143 (2000)
14. Regev, A., Panina, E.M., Silverman, W., Cardelli, L., Shapiro, E.: BioAmbients: an abstraction for biological compartments. Theor. Comput. Sci. **325**(1), 141–167 (2004)
15. Cardelli, L.: Brane calculi. In: Danos, V., Schachter, V. (eds.) CMSB 2004. LNCS, vol. 3082, pp. 257–278. Springer, Heidelberg (2005). https://doi.org/10.1007/978-3-540-25974-9_24
16. Fellermann, H., Cardelli, L.: Programming chemistry in DNA-addressable bioreactors. J. R. Soc. Interface **11**(99), 20130987 (2014)
17. Schöning, U.: Theoretische Informatik - kurz gefasst. Spektrum Akademischer Verlag (2003)
18. Weyland, M.S., et al.: The MATCHIT automaton: exploiting compartmentalization for the synthesis of branched polymers. Comput. Math. Methods Med. **2013**, 467428 (2013)
19. Gillespie, D.T.: A general method for numerically simulating the stochastic time evolution of coupled chemical reactions. J. Comput. Phys. **22**(4), 403–434 (1976)
20. Gillespie, D.T.: Exact stochastic simulation of coupled chemical reactions. J. Phys. Chem. **81**(25), 2340–2361 (1977)

Studying and Simulating
the Three-Dimensional Arrangement
of Droplets

Johannes Josef Schneider[1(✉)], Mathias Sebastian Weyland[1], Dandolo Flumini[1],
Hans-Georg Matuttis[2], Ingo Morgenstern[3], and Rudolf Marcel Füchslin[1,4]

[1] Institute for Applied Mathematics and Physics, Zurich University of Applied
Sciences, Technikumstr. 9, 8401 Winterthur, Switzerland
johannesjosefschneider@googlemail.com
{scnj,weyl,flum,furu}@zhaw.ch
[2] Department of Mechanical Engineering and Intelligent Systems, The University
of Electrocommunications, Chofu Chofugaoka 1-5-1, Tokyo 182-8585, Japan
hg@mce.uec.ac.jp
[3] Faculty of Physics, University of Regensburg, Universitätsstr. 31,
93053 Regensburg, Germany
ingo4004@aim.com
[4] European Centre for Living Technology, S. Marco 2940, 30124 Venice, Italy

Abstract. We present some work in progress on the development of a
probabilistic chemical compiler, being able to make a plan of how to create a three-dimensional agglomeration of artificial hierarchical cellular
constructs. These programmable discrete units offer a wide variety of
technical innovations, like a portable biochemical laboratory being able
to e.g. produce macromolecular medicine on demand, and of scientific
investigations, like contributions to questions regarding the origin of life.
This paper focuses on one specific issue of developing such a compiler,
namely the problem of simulating the experimentally observed spatial
transition from an originally one-dimensional lineup of droplets into a
three-dimensional, almost spherical arrangement, in which the droplets
form a network via bilayers connecting them and in which they are contained within some outer hull. The network created by the bilayers allows
the droplets to "communicate" (like agents in a multi agent system)
with each other and to exchange chemicals contained within them, thus
enabling a complex successive biochemical reaction scheme.

Keywords: Microfluidics · Droplet agglomeration · Monte Carlo

Supported by the European Horizon 2020 project *ACDC – Artificial Cells with Distributed Cores to Decipher Protein Function* under project number 824060.
The original version of this chapter was previously published non-open access. A Correction to this chapter is available at https://doi.org/10.1007/978-3-030-45016-8_18

F. Cicirelli et al. (Eds.): WIVACE 2019, CCIS 1200, pp. 158–170, 2020.
https://doi.org/10.1007/978-3-030-45016-8_16

1   Introduction

Over the last decades, huge progress has been made in biochemistry. A large amount of knowledge about the constituents and the processes within a cell has been gathered [1]. Even a new research field of synthetic biology has evolved [2], in which natural objects like the DNA in cells are purposedly altered or replaced in order to achieve some desired outcome, like producing a specific drug. Still, some questions remain unanswered so far, like one of the basic questions for the origin of life: Which constituent of a cell came first, the RNA or the cell membrane? Another problem turning up when considering the anthropic principle in cosmology is the discrepancy between the age of the earth and the time which would be needed by an ungoverned random evolutionary process to allow for the existence of higher-developed beings like humans.

In our approach, which we intend to follow within the European Horizon 2020 project *ACDC – Artificial Cells with Distributed Cores to Decipher Protein Function*, we do not consider fully equipped cells but the most simplified cell-like structures, being droplets comprised of some fluid, containing some chemicals, and surrounded by another fluid. As an additional feature, we also allow for droplets being contained within some outer hulls, playing the role membranes have for cells. These droplets arrange themselves in a three-dimensional way. Neighboring droplets, whose midpoints have a smaller distance to each other than the sum of their original radius values, can form bilayers between each other. Chemicals contained within the droplets can move to neighboring droplets through pores within these bilayers. Thus, a complex bilayer network is created, with the droplets being the nodes of this graph and the existing bilayers being represented by edges between the corresponding droplets. This bilayer network allows a controlled successive biochemical reaction scheme, leading to the intended macromolecules.

We aim at developing a probabilistic chemical compiler for a portable biochemical mini-laboratory, in which various desired macromolecules, like personalized antibiotics, can be produced on demand. Besides, we want to use this approach in order to determine up to which complexity higher-order macromolecules can be created by random agglomerations of droplets in order to make some contributions to Alexander Oparin's origin of life theory [3]. In small droplets, some metastable intermediate compounds can survive with larger probability than in the primordial soup, such that the generation of some macromolecules becomes more likely. Consequently, the time needed according to this theory for the development of life might be strongly reduced.

Summarizing, in the final stage of the project, our compiler for this biochemical device shall be able to solve the task to provide a recipe for producing the desired macromolecule, i.e., be able

- to determine the chemicals needed for the production of the macromolecule,
- to determine the gradual reaction steps leading finally to the desired macromolecule, with each reaction step being performed through a pore at the corresponding bilayer,

– to determine the bilayer network needed allowing for this reaction scheme, with the nodes of this network being the droplets filled with the chemicals determined in the first step and the edges being the bilayers, and finally
– to determine the experimental setup and parameters, leading to just this desired bilayer network.

Of course, in order to get there, our research has to start at the last item, in order to answer the question of how to design an experiment for generating and agglomerating droplets in a desired way.

2   Droplet Generation

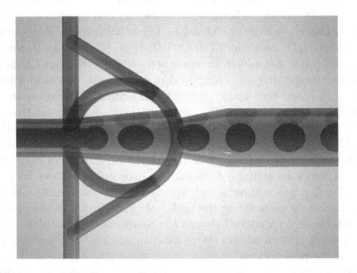

Fig. 1. Sketch of a so-called T-junction (schematically redrawn from [4]): the stream of the inner fluid is broken up under appropriate conditions. Spherical droplets are produced in the so-called dripping regime, in which the pressure within the fluid is neither too small nor too large.

Droplet generation, especially in the field of microfluidics, has been extensively studied over the past years [4–11]. A stream of fluid is broken up into droplets within a T-junction or some other antechamber, as shown in Fig. 1. The breaking-up of the stream is due to the fact that the shape of spherical droplets is energetically favorable when compared to a continuous stream of fluid under specific pressure conditions. The size of the droplets can be controlled by the respective flow rates. Due to the development of 3D printing technologies, producing such antechambers has become much easier and cheaper. Indeed, 3D printing has become a widely used technique in the field of microfluidics [12–19]. We thus consider the general problem of producing droplets to be solved, except that the compiler has to choose appropriate antechambers for the production process or even to create them using the 3D printing technology.

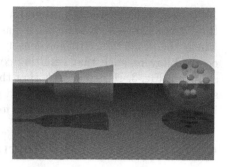

Fig. 2. Sketch of the initial and final states of the spatial rearrangement of droplets

3   3D Rearrangement of Droplets

In the experiments carried out by Jin Li, member of our collaborating group of David A. Barrow at Cardiff University, the droplets leave the antechamber and first enter a widening tube, in which they e.g. form a zig-zag line. Then they move on to an expansion chamber, in which they rearrange themselves in a three-dimensional way, where several of them are surrounded by some newly generated hull [4], as shown in Fig. 2. Our first main task in this problem will be to study and understand this rearrangement process at least so far that we can simulate it to obtain the same types of three-dimensional arrangements of droplets as found in experiments. We will of course never be able to reproduce the experiments exactly, first of all, because not all experimental parameters are known, secondly also because of lack of computing time.

Fig. 3. Resulting 3D arrangement containing bilayers: These two graphics show the same configuration, but on the right, the size of the droplets is reduced in order to visualize the bilayer network.

In some experiments, the resulting three-dimensional arrangements indeed look like in Fig. 2: Jin Li managed to create configurations with up to roughly

100 small droplets swirling around within an outer hull with a diameter smaller than 1 cm [20]. We, on the other hand, are interested in those experiments in which the droplets are rather densely packed and in which they are enabled to create bilayers, such that resulting configurations look more similar to the one shown in Fig. 3. Here one finds that the droplets more or less lose their spherical shapes due to bilayer formation and that a complex graph is formed by these bilayers which shall be used later on for the successive chemical reaction scheme.

4   Limitations of the Rearrangement Process

This rearrangement process does not allow for the formation of any desired bilayer network, instead, it is restricted by mathematical and physical limitations. The surface tension of the droplets is rather high compared with the other forces (adhesion, inertial forces due to changes of the flow field) in the system, so the shape of the droplets remains more or less close to spherical. Volume changes can be neglected, as the velocity range is limited to incompressible flow, far below 1/10 of the sound velocity of the fluids involved. Therefore, as starting points of our discussion of examples for restrictions for the arrangement of the droplets, the exact solutions of packing problems for rigid spheres are suitable, and we will augment our discussions by taking into account deformations and bounding layers in the next step where necessary.

Fig. 4. Kissing number problem in three dimensions: The maximum number of spheres touching a sphere in their midst is 12. This configuration also resembles the optimum packing of 13 spheres in a sphere in three dimensions.

The most prominent of these examples limiting the types of achievable bilayer networks is related to the kissing number problem. The kissing number problem is stated as follows: given equal spheres of the same size in D dimensions, what is the maximum number of spheres being able to touch a sphere in their midst? While this problem is trivial to solve in one and two dimensions, for which the kissing number is 2 and 6, rsp., it led to a famous dispute between David Gregory and Isaac Newton at the end of the 17th century, whose details however are still under debate [21]. According to the most widely accepted version of the anecdote, Newton proposed that the number was 12, while Gregory argued that the number had to be 13, as a further sphere could be placed close enough to also touch the sphere in the midst. It took till the 1950s to prove that Newton had been right [22], the space was just not sufficient to allow for a 13th sphere to touch the central sphere. Thus, if the droplets keep their spherical shapes, one droplet can only be touched by up to 12 other droplets of the same size. However, if the droplets lose their spherical shapes due to bilayer formation, 13 or even slightly larger numbers of touching droplets can be obtained.

 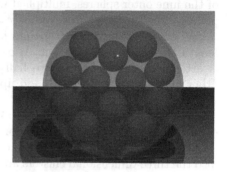

Fig. 5. One-dimensional sausage and two-dimensional pizza configuration for 13 spheres: The configuration on the right resembles the densest packing of 13 circles in a circle.

The next example to be considered here is related to the problem of gaining the densest configuration. If assuming that the volume enclosed in the convex outer hull is to be minimized, the question arises why the droplets rearrange themselves in a three-dimensional way at all. One wonders why they do not stay in a one-dimensional lineup, which is also called "sausage", or form a two-dimensional configuration, called "pizza" [23], as shown in Fig. 5. While it is trivial to determine the optimum one-dimensional lineup, the two-dimensional arrangement is derived from the optimum packing for N circles of equal radius r within a circumcircle of minimum radius R. The optimum value for the radius of the circumcircle of this two-dimensional arrangement was proven to be $R_{2D} = (2 + \sqrt{5})r$ [24] for $N = 13$. Then the convex outer hull does of course not need to be spherical, as depicted in Fig. 5, but it could enclose e.g. the one-dimensional lineup in a cylindric way with two half-spheres attached to the ends of the cylinder. For $N = 13$ spheres of radius r, the enclosed volume of this one-dimensional lineup would only have to be

$V_{1D} = \pi r^2 \times 2r(N - 1) + 4\pi/3r^3 = 25\frac{1}{3}r^3\pi$, while for the corresponding three-dimensional configuration shown in Fig. 4, the enclosed volume would almost have to be $V_{3D} \lesssim 4\pi/3(3r)^3 = 36r^3\pi$. (Please note that the minimum value for the volume of the three-dimensional arrangement is slightly smaller than this value, as some small parts of the surrounding spherical hull can be cut off because surfaces over triangles of spheres can be made partially planar, while the hull still remains convex.) Only at larger numbers N of spheres, like $N = 56$ [25], a three-dimensional cluster is more densely packed than the one-dimensional sausage. This transition, which is also called sausage catastrophe, is still under research debate. According to the sausage assumption, intermediate dimensional structures like pizzas are never optimum. This can also be seen for our exemplary two-dimensional pizza configuration shown in the right half of Fig. 5: the four inner spheres can be neglected when calculating the volume V_{2D} of the convex surrounding hull of this configuration. V_{2D} consists of three parts: the inner part is given by the area A within the polygon, formed by the midpoints of the nine outer spheres, multiplied with the height $2r$. To each of the nine side planes, a half cylinder with radius r and a length corresponding to the length of the edge between the midpoints is attached. These lengths sum up to the length U of the closed polygon. At each node of the polygon, a spherical wedge is attached, connecting the two half cylinders ending at that node with each other. These spherical wedges add up to a sphere with radius r. Summarizing, we get $V_{2D} = A \times 2r + U \times r^2\pi/2 + 4\pi r^3/3 \sim 29.6 \times 2r^3 + 19.8 \times r^3\pi/2 + 4\pi r^3/3 \sim 94.5r^3$. Therefore, we have $V_{1D} < V_{2D} < V_{3D}$. Thus, the minimization of the volume within a hull has no dominating effect on the arrangement process, on the contrary, one even finds in configurations resulting in experiments that also the three-dimensional clusters are not most densely packed. They sometimes even contain holes in which a further droplet could be placed [4].

Another picture is obtained if we aim at minimizing the surface of the surrounding hull. If we again have a look at our example with 13 spheres, we find $S_{1D} = 12 \times 2r \times 2r\pi + 2 \times 2r^2\pi = 52\pi r^2 \sim 163r^2$ for the surface of the one-dimensional sausage, $S_{2D} = 2 \times A + \pi r \times U + 4\pi r^2 \sim 134r^2$ for the surface of the two-dimensional pizza, and $S_{3D} \lesssim 4\pi(3r)^2 = 36\pi r^2 \sim 113r^2$ in the case of three dimensions. Summarizing, we find that $S_{3D} < S_{2D} < S_{1D}$, i.e., the minimization of the surface of the hull or, physically speaking, the surface tension could have a large effect on the agglomeration process of the droplets. However, the minimization of the surface does not totally dominate this process, as the resulting shapes of the hulls as seen in the videos generated by Jin Li show perfectly spherical or elliptical shapes or sometimes oval shapes due to boundaries, but never shapes with triangular planes. But one must not forget that on the one hand not only droplets but also some fluid around them is contained within the outer hull and that on the other hand the surface tension tends to minimize local deviations from the average curvature radius.

Finally, we want to deal with the quest for the one and only central sphere. The so-called ideal picture which is often drawn on blackboards depicts a central sphere being surrounded by some number $N - 1$ of other spheres touching it and

Fig. 6. Configuration consisting of five spheres with their centers being placed on the corners of a regular pentagon touching their neighbors and a sixth sphere touching all other spheres

touching their neighbors, as shown in Fig. 6. This picture is motivated by its two-dimensional analogon, in which six circles can be placed around a seventh circle. However, this picture with a central circle is only valid for $N \leq 9$ in two dimensions, as there are at least two inner circles in optimum packings of circles for $N \geq 10$. Transferring this picture to three dimensions by replacing the circles with spheres, the situation becomes unstable for $N \geq 7$ spheres, as the sphere in the center can move freely in the third dimension and would thus fall through the ring, due to the law of gravity. The densest packing of seven spheres within a spherical hull is obtained for $R \sim 2.59r$ but it does not contain a sphere which could be classified as center sphere [26]. Stable configurations with $N - 1$ spheres being placed on a regular $N - 1$-gon and touching their neighbors and an Nth sphere touching all the others can only be obtained for $N = 4, 5$, and 6, under the condition that the radius R of the outer hull has a specific value, such that the Nth sphere at the bottom of the configuration neither drops down nor presses the other spheres apart such that the connections between them are destroyed:

If placing three spheres on the edges of an equilateral triangle with side length $2r$ and a fourth sphere centered below them, touching the other three spheres, one gets the densest packing of four spheres in a sphere, which remains stable in a spherical hull with radius $R = (1 + \sqrt{6}/2)r \sim 2.22r$.

If placing four spheres on the edges of a square with side length $2r$ and a fifth sphere centered below them, touching the other four spheres, one gets the densest packing of five spheres in a sphere [26]. It remains stable within a spherical hull with radius $R = (1 + \sqrt{2})r \sim 2.41r$. One can even place a sixth sphere symmetrically to the fifth sphere on the opposite side of the center square, thus achieving the densest packing of six spheres in a sphere.

If placing five spheres on the edges of a regular pentagon with side length $2r$ and then a sixth sphere below them, touching all of them, as shown in Fig. 6, one has to make a larger effort to determine a radius R of the spherical surrounding hull to get a stable configuration. The radius of the pentagon formed by the midpoints of the spheres can be easily determined as $r\sqrt{2/(1 - \cos(2\pi/5))} \sim 1.7r$ geometrically or as $r\sqrt{50 + 10\sqrt{5}}/5$ making use of the golden ratio. If imagining the pentagon formed by the midpoints of the five spheres being placed in the xy-plane centered around the origin, one can place the sixth sphere, which is supposed to touch all the other spheres, on the z-axis at $z_6 = -r\sqrt{4 - 2/(1 - \cos(2\pi/5))} \sim -1.05r$. In order to get a stable configuration in which this sixth sphere does not drop down, we need to blow up the radius of the surrounding hull to $R = r - 2r^2/z_6 - z_6 \sim 3.95r$, its midpoint lies at $z_M = -2r^2/z_6$.

Thus, mathematically speaking, for each of these scenarios, there is only one exact value R for the radius of the surrounding spherical hull, for which the inner configuration of hard spheres is stable. But, of course, if allowing some amount of deformation of the spheres and also of the hull, one gets a range of possible radius values instead of one exact sharp value only.

5   Simulating the Rearrangement Process

After these initial considerations, we now describe how we intend to simulate the rearrangement process. We will perform macroscale Monte Carlo movement simulations, imitating the movement behavior of the droplets being first lined up in an almost one-dimensional structure within the T-junction or some other antechamber and then entering the expansion chamber, in which they rearrange themselves in a three-dimensional way within some outer hull, as shown schematically in Fig. 2. During this rearrangement process, some droplets touching each other will form bilayers [27]. These bilayers can be reshaped, broken up, and newly formed, depending on the stability of the bilayers [28]. When bilayers are created, the droplets lose their spherical shape. We will test various ways to simulate the formation, change, and destruction of bilayers and the change of the shape of the cores in a computationally not too expensive way. A cheap way would be to place the particles on a regular or irregular lattice [29,30] and even to make use of a cellular automata approach as in traffic dynamics [31,32], but this approach is not feasible as it restricts the possibilities for resulting bilayer configurations too much. We intend to invent an entirely new method of Monte Carlo movement simulations of such droplets, as existing methods like in [33] put too much emphasis on the resulting network of droplets, introducing springs between these particles already from the very beginning, while these droplets move rather independently of each other at first in the experiments, as seen in movies generated by Jin Li [20]. Only at a later stage when they are already surrounded by some hull, the droplets gradually settle down, reducing their individual behaviors, and start to move coherently. While the specific spatial setup of an experiment with preset values for widths and lengths of various parts of the junction can be easily employed also in the Monte Carlo simulation, it is a harder task to find appropriate values for the probabilities for braking and acceleration of droplets as well as for bilayer formation and destruction and also for some introduction of random movement. These values depend on various experimental parameters, like pressure and viscosity. We intend to adjust the parameters for the Monte Carlo simulation in a way that the resulting configurations reflect the three-dimensional arrangements of droplets as found in experiments.

In order to perform fast simulations with our limited computing time, we will at first consider spherical droplets only. If two droplets form a bilayer with each other, we will compensate the overlapping volumes by increasing the radii of the corresponding droplets. Furthermore, we will either select particles in random sequential order and ask the process which move to perform with them (but then some larger gaps in the configurations could occur and only be slowly resolved) or we start off with the outermost droplet at the right (if they are moving from left to right), then the second rightest droplet, and so on, until at last the outermost left droplet is chosen. Furthermore, we retain the information whether a droplet is connected to other droplets via bilayers, such that they can move with each other, or already part of a group of droplets, with which it moves more and more coherently within an outer hull. In this case, all droplets within such a group are simultaneously updated. The movement of the various droplets within a group is split in a movement process for the center of mass of that group and a movement process for the specific droplet relative to the center of mass.

Summarizing, we need to implement the following processes in the Monte Carlo movement simulation:

- acceleration process: With some probability, a droplet is accelerated until it reaches its desired velocity.
- braking process: If walls or other droplets provide obstacles for the movement of a droplet, it of course has to brake. Otherwise, there is also some probability for braking.
- random movement process: We will to some extent also allow random movement, i.e., the velocity vector can be slightly altered.
- bilayer formation process: With some probability, droplets touching (or in the simulation even overlapping) each other can form bilayers. The probability for bilayer formation increases with increasing overlap and increasing time for which the overlap already lasted.
- bilayer destruction process: With some probability, a bilayer formed can also be destroyed again. This probability could depend on the length of time for which the bilayer was in existence.

Related to these processes, we thus have the probabilities $p_{accelerate}$, p_{brake}, $p_{randommovement}$, $p_{bilayerformation}$, and $p_{bilayerdestruction}$. Thus, our simulations will not contain experimental parameters like viscosity in an explicit way, but we will find that implicitly e.g. the probabilities for braking and acceleration will depend on viscosity and other parameters, such that one task will be to find an appropriate mapping between experimental input values and parameters in the computer simulations.

6 Summary

In this paper, we described the first step to be undertaken in the development of a chemical compiler for a biochemical micro-laboratory device, with which e.g. macromolecules shall be produced on demand. For this device, we want to make use of the ACDC technology, i.e., systems of droplets which agglomerate in a three-dimensional way, forming bilayers between them. This bilayer network will allow for a step-wise generation of some desired macromolecules, which are gradually constructed from smaller units, being contained in the various droplets, with the successive chemical reactions being enabled via the bilayers formed between neighboring droplets. Such a compiler has been exemplarily already developed for one specific molecule [34]. In this project, this compiler has to be generalized and also made probabilistic because of the variability in the rearrangement process. When performing simulations for the three-dimensional rearrangement of droplets as seen in experiments, our objective is not to find e.g. the densest configuration possible [35,36], but to find configurations most similar to those resulting in experiments.

Acknowledgment. Fruitful discussions with Jin Li, David A. Barrow, and Oliver Castell at the Cardiff University in Wales and with Marco Baioletti, Marco Villani, and Roberto Serra at the Wivace 2019 conference at the Università della Calabria in Rende, Italy are kindly acknowledged. This work has been financially supported by the European Horizon 2020 project *ACDC – Artificial Cells with Distributed Cores to Decipher Protein Function* under project number 824060.

References

1. Alberts, B., et al.: Molecular Biology of The Cell. Garland Science, 6th edn. Taylor & Francis, New York (2014)
2. Gibson, D.G., Hutchison III, C.A., Smith, H.O., Venter, J.C. (eds.): Synthetic Biology - Tools for Engineering Biological Systems. Cold Spring Harbor. Cold Spring Harbor Laboratory Press, New York (2017)
3. Oparin, A.I.: The Origin of Life on the Earth, 3rd edn. Academic Press, New York (1957)
4. Li, J., Barrow, D.A.: A new droplet-forming fluidic junction for the generation of highly compartmentalised capsules. Lab Chip **17**, 2873–2881 (2017)
5. Morgan, A.J.L., et al.: Simple and versatile 3D printed microfluidics using fused filament fabrication. PLoS ONE **11**(4), e0152023 (2016)
6. Eggers, J.: Nonlinear dynamics and breakup of free-surface flows. Rev. Mod. Phys. **69**, 865 (1997)
7. Eggers, J., Villermaux, E.: Physics of liquid jets. Rep. Progress Phys. **71**, 036601 (2008)

8. Link, D.R., Anna, S.L., Weitz, D.A., Stone, H.A.: Geometrically mediated breakup of drops in microfluidic devices. Phys. Rev. Lett. **92**, 054503 (2004)
9. Garstecki, P., Stone, H.A., Whitesides, G.M.: Mechanism for flow-rate controlled breakup in confined geometries: a route to monodisperse emulsions. Phys. Rev. Lett. **94**, 164501 (2005)
10. Garstecki, P., Fuerstman, M.J., Stone, H.A., Whitesides, G.M.: Formation of droplets and bubbles in a microfluidic T-junction - scaling and mechanism of breakup. Lab Chip **6**, 437–446 (2006)
11. Guillot, P., Colin, A., Ajdari, A.: Stability of a jet in confined pressure-driven biphasic flows at low Reynolds number in various geometries. Phys. Rev. E **78**, 016307 (2008)
12. Au, A.K., Huynh, W., Horowitz, L.F., Folch, A.: 3D-printed microfluidics. Angew. Chem. Int. Ed. **55**, 3862–3881 (2016)
13. Takenaga, S., et al.: Fabrication of biocompatible lab-on-chip devices for biomedical applications by means of a 3D-printing process. Phys. Status Solidi **212**, 1347–1352 (2015)
14. Macdonald, N.P., Cabot, J.M., Smejkal, P., Guit, R.M., Paull, B., Breadmore, M.C.: Comparing microfluidic performance of three-dimensional (3D) printing platforms. Anal. Chem. **89**, 3858–3866 (2017)
15. Lee, K.G., et al.: 3D printed modules for integrated microfluidic devices. RSC Adv. **4**, 32876–32880 (2014)
16. Yazdi, A.A., Popma, A., Wong, W., Nguyen, T., Pan, Y., Xu, J.: 3D Printing: an emerging tool for novel microfluidics and lab-on-a-chip applications. Microfluid. Nanofluid **20**, 1–18 (2016)
17. Chen, C., Mehl, B.T., Munshi, A.S., Townsend, A.D., Spence, D.M., Martin, R.S.: 3D-printed microfluidic devices: fabrication, advantages and limitations - a mini review. Anal. Methods **8**, 6005–6012 (2016)
18. He, Y., Wu, Y., Fu, J., Gao, Q., Qiu, J.: Developments of 3D printing microfluidics and applications in chemistry and biology: a review. Electroanalysis **28**, 1–22 (2016)
19. Tasoglu, S., Folch, A. (eds.): 3D Printed Microfluidic Devices. MDPI, Basel (2018)
20. Li, J.: Private communication (2019)
21. Casselman, B.: The difficulties of kissing in three dimensions. Notes AMS **51**, 884–885 (2004)
22. Schütte, K., van der Waerden, B.L.: Das problem der 13 Kugeln. Math. Annalen **125**, 325–334 (1953)
23. Wills, J.: Spheres and sausages, crystal and catastrophes - and a joint packing theory. Math. Intelligencer **20**, 16–21 (1998)
24. Fodor, F.: The densest packing of 13 Congruent circles in a circle. Contributions Algebra Geometry **44**, 431–440 (2003)
25. Gandini, P.M., Wills, J.M.: On finite sphere-packings. Math. Pann. **3**, 19–20 (1992)
26. https://en.wikipedia.org/wiki/Sphere_packing_in_a_sphere. Accessed 27 Dec 2019
27. Mruetusatorn, P., Boreyko, J.B., Venkatesan, G.A., Sarles, S.A., Hayes, D.G., Collier, C.P.: Dynamic morphologies of microscale droplet interface bilayers. Soft Matter **10**, 2530–2538 (2014)
28. Guiselin, B., Law, J.O., Chakrabarti, B., Kusumaatmaja, H.: Dynamic morphologies and stability of droplet interface bilayers. Phys. Rev. Lett. **120**, 238001 (2018)
29. Kirkpatrick, S., Schneider, J.J.: How smart does an agent need to be? Int. J. Mod. Phys. C **16**, 139–155 (2005)
30. Schneider, J.J., Hirtreiter, C.: The democracy-ochlocracy-dictatorship transition in the Sznajd model and in the Ising model. Physica A **353**, 539–554 (2005)

31. Nagel, K., Schreckenberg, M.: A cellular automaton model for freeway traffic. J. Phys. I France **2**, 2221–2229 (1992)
32. Schneider, J., Ebersbach, A.: Anticipatory drivers in the nagel-schreckenberg-model. Int. J. Mod. Phys. C **13**, 107–113 (2002)
33. Griesbauer, J., Seeger, H., Wixforth, A., Schneider, M.F.: Method for the Monte Carlo Simulation of Lipid Monolayers including Lipid Movement. https://arxiv.org/pdf/1012.4973.pdf (2010). Accessed 27 Dec 2019
34. Weyland, M.S., et al.: The MATCHIT automaton: exploiting compartmentalization for the synthesis of branched polymers. Comp. Math. Meth. Med. **2013**, 467428 (2013)
35. Müller, A., Schneider, J.J., Schömer, E.: Packing a multidisperse system of hard disks in a circular environment. Phys. Rev. E **79**, 021102 (2009)
36. Schneider, J.J., Müller, A., Schömer, E.: Ultrametricity property of energy landscapes of multidisperse packing problems. Phys. Rev. E **79**, 031122 (2009)

Investigating Three-Dimensional Arrangements of Droplets

Johannes Josef Schneider[1(✉)], Mathias Sebastian Weyland[1],
Dandolo Flumini[1], and Rudolf Marcel Füchslin[1,2]

[1] Institute for Applied Mathematics and Physics, Zurich University of Applied
Sciences, Technikumstr. 9, 8401 Winterthur, Switzerland
johannesjosefschneider@googlemail.com
{scnj,weyl,flum,furu}@zhaw.ch
[2] European Centre for Living Technology, S. Marco 2940, 30124 Venice, Italy

Abstract. We present some work in progress on the development of
a small portable biochemical laboratory, in which spatially structured
chemical reaction chains in a microfluidic setting shall be created on
demand. For this purpose, hierarchical three-dimensional agglomerations
of artificial cellular constructs are generated which will allow for a governed gradual reaction scheme leading e.g. to desired macromolecules. In
this paper, we focus on the task of investigating the bilayer networks via
which the chemical reactions are performed, both from experiment and
from simulation.

Keywords: Microfluidics · Replica symmetry breaking ·
Ultrametricity

1 Introduction

Within the European Horizon 2020 project *ACDC – Artificial Cells with Distributed Cores to Decipher Protein Function*, our group aims at the development
of a cheap and small portable biochemical laboratory being able to e.g. produce any desired macromolecule, like antibiotics, on demand. For this purpose,
biologically inspired structures, namely droplets comprised of some fluid and
being surrounded by another fluid, shall be used. The droplets may also contain
chemicals and can be enclosed within some outer hulls, mimicking the role membranes have for cells. Neighboring droplets can form bilayers and can exchange
the chemicals within them through pores in these bilayers, thus allowing for
chemical reactions. A specific three-dimensional arrangement of droplets and a

Supported by the European Horizon 2020 project *ACDC – Artificial Cells with Distributed Cores to Decipher Protein Function* under project number 824060.
The original version of this chapter was previously published non-open access. A Correction to this chapter is available at https://doi.org/10.1007/978-3-030-45016-8_18

F. Cicirelli et al. (Eds.): WIVACE 2019, CCIS 1200, pp. 171–184, 2020.
https://doi.org/10.1007/978-3-030-45016-8_17

resulting bilayer network is then intended to produce a desired chain of reactions, e.g. resulting in a macromolecule via a gradual biochemical reaction scheme. In order to govern the process of creating a specific three-dimensional arrangement of droplets within the biochemical laboratory, a control unit is needed which is termed 'chemical compiler' by us. This chemical compiler must be able to define with which chemicals the various droplets need to be filled and to determine which bilayer networks could be used for the desired biochemical reaction chain. It also needs to foretell how to create a three-dimensional arrangement of droplets from the originally one-dimensional lineup of droplets [1], leading to one of the required bilayer networks. For the development of such a compiler, still a lot of research has to be invested. We already proved that the existence of such a compiler is possible, both by showing for an example that it can be created [2] and by providing a general mapping from a computer program written only with the commands `if-then-else` and `goto` onto a biochemical reaction scheme [3]. But we also already showed that not any bilayer network desired for a specific biochemical reaction chain can be created, due to physical and mathematical limitations to achievable three-dimensional arrangements of droplets [4].

In order to finally be able to build such a device and to control it by means of such a compiler, we have to get a better understanding of the consequences of experimental settings and other input parameters of experiments, of the experimental progress, and also of the outcome of the experiments. The process of creating droplets is in the meantime very well understood and can be governed at will [1,5–11], especially by the usage of 3D printing technology, which has become a widely used technique in the field of microfluidics [12–19]. We thus consider the general problem of producing droplets to be solved, except that the compiler has to choose appropriate antechambers for the production process or even to create them using the 3D printing technology.

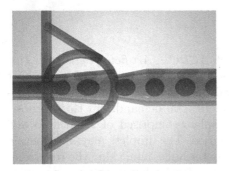 

Fig. 1. Sketch of the initial and final states of the spatial rearrangement of droplets.

However, the process of creating three-dimensional arrangements of droplets from original one-dimensional lineups, as shown in Fig. 1, is not yet fully understood and is still a current research project. Our collaborating experimental group at Cardiff University currently extends the boundaries of this field, now being able to include more than 100 droplets swirling within a droplet of a diameter of roughly 1 cm [20]. Thus, in one part of our project, we want to understand how this transition works and to perform Monte Carlo simulations mimicking the experimental setup, as described in [4]. We will of course never be able to accurately reproduce the overall experiment in our simulations, due to lack both of computing time and of information about experimental parameters. However, we intend to adjust the parameters for the simulation in a way that the resulting configurations reflect the three-dimensional arrangements of droplets as found in experiments. But of equal importance is the question of the outcoming three-dimensional configurations, which are not always the same, as seen in movies generated from experiments [20]. Instead of a one-and-only final configuration, various arrangements are achieved. But when looking closely at the resulting configurations, one finds that they are not entirely random and are often rather similar to each other and that maybe even some configurations might be identical.

As already mentioned, we will finally have to rely on the existence of bilayer networks allowing for the complex gradual reaction scheme in order to produce the desired reaction structures. Therefore, we have the task to closely investigate the resulting three-dimensional configurations both from experiments and from simulations, which proves to be a very difficult problem, as described in the next sections. In Sect. 2, we start off with describing when we consider two differently looking configurations to be identical. In Sect. 3, we continue this consideration by describing how to detect that two configurations are indeed identical. But as already mentioned, the resulting configurations from experiments are not always identical, such that we continue with various approaches dealing with differences among them. If there are only small differences among various configurations, which are caused by single droplets that might be differently connected and placed, then searching for a network core might be the suitable strategy as mentioned in Sect. 4. But if the resulting configurations are more than only slightly different, a strategy called Searching for Backbones detecting network parts common to all configurations, which are called backbones, has to be applied as described in Sect. 5. However, there could be so many differences among the resulting configurations that not even a single larger backbone can be found. But still, there could be various groups of configurations with only little differences among the configurations within each group and larger differences between configurations in different groups. Sections 6 and 7 provide two approaches for detecting such groups and ways how to merge them into even larger supergroups.

2 Comparing Resulting Configurations

Two exemplary three-dimensional arrangements of droplets are shown in the upper half of Fig. 2. At first sight, they look different, but this might be the case for any two three-dimensional spherical objects consisting of various constituents. Usually, one would claim that two such three-dimensional objects are

Fig. 2. Top: Two resulting three-dimensional arrangements of droplets filled with various chemicals. Bottom: Corresponding bilayer networks between the droplets. We made these bilayer networks visible by reducing the size of the droplets and printing a connecting edge between a pair of neighboring droplets if they have formed a bilayer.

identical if they can be made congruent by rotation and mirroring. However, in our problem, the various droplets can be more or less deformed, such that the congruency method has to fail. And we do not need to care about what the overall configuration exactly looks like, all we need to care about is the underlying bilayer network formed by the bilayers between pairs of droplets, allowing for or denying the complex gradual reaction scheme we have in mind.

Concluding, for our problem, two three-dimensional arrangements are identical if their underlying bilayer networks are identical, because due to their identity, they lead to identical possibilities for chemical reactions and thus to identical reaction results.

3   The Graph Isomorphism Problem

Thus, the question arises how to show that two such bilayer networks are identical. At first, one might think that it must be easy to see that two such networks are identical, but this is not the case, as we want to demonstrate at a simple example shown in Fig. 3. This example consists of two graphs (Throughout this paper, we will use both the term network, which is used in physics

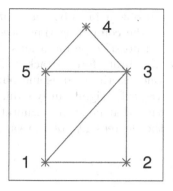
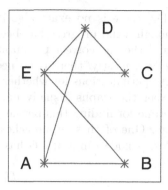

Fig. 3. Simple example for the graph isomorphism problem: The two graphs shown are not identical at first sight. But using the isomorphism $E = 3, A = 5, B = 4, D = 1$, and $C = 2$, one finds that these two graphs can be transferred one into the other by this isomorphism and can thus be considered to be identical.

and biology, and the term graph, which is used by mathematicians. Both terms describe the same.) with five nodes and seven edges each. At first sight, these two graphs seem to be different. This seemingly difference could be caused by e.g. rotating and mirroring the droplet configuration, such that we could be misled by eyesight. In order to avoid this effect, usually, the adjacency matrix η_G of graph G is considered with

$$\eta_G(i,j) = \begin{cases} 1 & \text{if nodes } i \text{ and } j \text{ are connected} \\ & \text{by an edge in graph } G \\ 0 & \text{otherwise} \end{cases} \tag{1}$$

For the left graph (lg) and the right graph (rg) in Fig. 3, we thus get the corresponding adjacency matrices

$$\eta_{\text{lg}} = \begin{pmatrix} 0 & 1 & 1 & 0 & 1 \\ 1 & 0 & 1 & 0 & 0 \\ 1 & 1 & 0 & 1 & 1 \\ 0 & 0 & 1 & 0 & 1 \\ 1 & 0 & 1 & 1 & 0 \end{pmatrix} \quad \text{and } \eta_{\text{rg}} = \begin{pmatrix} 0 & 1 & 0 & 1 & 1 \\ 1 & 0 & 0 & 0 & 1 \\ 0 & 0 & 0 & 1 & 1 \\ 1 & 0 & 1 & 0 & 1 \\ 1 & 1 & 1 & 1 & 0 \end{pmatrix}, \tag{2}$$

rsp., with the node orderings $1 - 2 - 3 - 4 - 5$ used for the left graph and $A - B - C - D - E$ used for the right graph.

As we see in Eq. (2), having a look at the adjacency matrices does not help much. Nevertheless, as already mentioned in the caption of Fig. 3, an isomorphism can be found, mapping the nodes of the right graph onto their corresponding counterparts in the left graph and thus showing that these two graphs are identical according to our definition.

Thus, the task is to find a permutation σ of the nodes in the way that

$$\eta_{\text{lg}}(\sigma(i), \sigma(j)) = \eta_{\text{rg}}(i,j) \quad \text{for all pairs of nodes}(i,j). \tag{3}$$

So far, there is no exact algorithm solving this problem in polynomial time, although some progress has been made to reduce the computing time needed to find the one correct permutation among the $N!$ possible permutations [21]. However, mostly, there is no need to make use of a brute force attack testing all $N!$ permutations for whether it is the correct one describing an isomorphism between the graphs. Usually, one can make some tricks for significantly reducing the time for finding this permutation or for showing that no such permutation exists. One of these approaches is to consider the degrees D_G of the various nodes. A node i in graph G has degree

$$D_G(i) = \sum_j \eta_G(i,j). \tag{4}$$

The degree of a node counts the number of edges attached to this node, i.e., it measures the number of nodes with which it is directly connected via an edge. For our example in Fig. 3, we thus have $D_{lg}(1) = 3$, $D_{lg}(2) = 2$, $D_{lg}(3) = 4$, $D_{lg}(4) = 2$, and $D_{lg}(5) = 3$ for the left graph and $D_{rg}(A) = 3$, $D_{rg}(B) = 2$, $D_{rg}(C) = 2$, $D_{rg}(D) = 3$, and $D_{rg}(E) = 4$ for the right graph. Thus, we already know at this point that the permutation σ, if it is to describe an isomorphism, has to map node 3 onto node E, as these are the only nodes with degree 4. Then we can either map node 1 onto node A and 5 onto node D or map node 1 onto node D and 5 onto node A. And analogously we have two possibilities for mapping the nodes with degree 2, we can either map node 2 onto node B and 4 onto node C or map node 2 onto node C and 4 onto node B. Thus, from the original number $5! = 120$ of possible permutations, only $2 \times 2 = 4$ possible permutations remain which have to be checked whether one of them provides the correct isomorphism.

For bilayer networks comprised of edges connecting droplets filled with different chemicals, the situation is even better. These droplets filled with different chemicals correspond in graph theory to so-called colored nodes. There is a whole bunch of famous problems dealing with colored graphs and graph coloring [22], e.g., the question how many colors are needed at least to color the countries on a map in the way that no two neighboring countries share the same color. While for a globe of the earth and for a map in an atlas, a minimum of four colors is required, seven colors would be needed for the corresponding problem on a torus [23]. If we add the requirement that no two nodes connected by an edge in a bilayer network have the same color, an even larger minimum number of colors could be needed, depending on the network topology.

For our problem of finding out whether two configurations are identical, i.e., whether the underlying bilayer networks can be mapped onto each other by an isomorphism σ, we only need to consider those permutations which map the nodes in a way that they stay with the same color, i.e., that they contain the same chemical molecules. Together with approaches like the degree measurement, the computation time needed for finding the isomorphism σ or for proving that the two configurations differ can be strongly reduced.

4   Searching for a Network Core

However, the situation is not so easy. As already mentioned, differences between resulting three-dimensional arrangements can be observed very often, such that it is not too likely that we will one day be able to govern the experiment in a way that all resulting three-dimensional arrangements of droplets share exactly the same underlying bilayer network.

Fig. 4. Simple example for four graphs sharing the same network core but also containing one node attached in various ways to other nodes

Instead, we will have to deal with different bilayer networks. In the simplest case, it will be one droplet or a small number of droplets leading to differences between the various resulting networks. As an example, four different possibilities are shown in Fig. 4. In this figure, the new node with No. 6 is located at (almost) the same place in the first three examples, either it is attached to one of the two neighboring nodes or it is even attached to both. But it could also be, as depicted in the fourth example, that such a node is sometimes located at one place and sometimes at another, thus being able to form bilayers with various nodes.

These graphs are of course not identical, no isomorphism exists to map one of these graphs onto another graph, except for the second and the fourth graph between which an isomorphism can be found. Thus, we at first sight seem to be unable to follow the approach described in the last section. Nevertheless, we see that these graphs are almost identical. They share some common core network. The question is how to find out whether networks only differ slightly and share a common core.

One simple approach to this problem of finding a network core was applied by Kirkpatrick and coworkers [24], who simply tried to cut off those network parts which would most likely cause these differences. The idea behind this approach is that the likeliest candidates would be those with minimum degree. For our example in Fig. 4, if we cut off all nodes with degree 1, we would mostly find that the resulting networks share the same core. Only the third example would seemingly have another core. But if we went on also to cut off all nodes with degree 2, then we would e.g. mostly also cut off node No. 2. This shows the limitations of this approach. Another approach suggested by Kirkpatrick would be to successively cut off all nodes with degree 1. Thus in the second iteration of this idea, one would only cut off those nodes which had degree 2 in the original network, but already lost one neighbor which was connected only to it. However, this approach would also not help for resolving the network core in the third example depicted in Fig. 4.

If assuming that the differences between various configurations are indeed so small and of a kind that a common network core exists and can be found, then one still faces the graph isomorphism problem again. However, here an isomorphism has to be found not for the whole graph but only for the assumed network core. If there are larger differences, other strategies, like Searching for Backbones, have to be used which focus on similarities.

5   Searching for Backbones

For various complex optimization problems, like the traveling salesman problem, the vehicle routing problem, and the problem of finding ground states for spin glasses, it has been found that different quasi-optimum configurations share entirely identical backbones [25–29], i.e., parts which are identical in all these configurations.

For finding this backbone summarizing the common parts of different configurations, an overlap measure $Q(\sigma, \tau)$ has to be defined. Due to the differences between the various configurations on the one side and the graph isomorphism problem on the other side, we cannot simply make use of the adjacency matrices as in [27] for our bilayer network problem. Instead, here an overlap measure $Q(\sigma, \tau)$ would need to count all common properties between the configurations σ and τ. For example, if both configurations share an edge between a red and a green node, then the overlap is incremented by 1. Of course, such an overlap measure gets maximal if a configuration is compared with itself. However, the various $Q(\sigma, \sigma)$ might have slightly different values in our problem, as this approach depends strongly on the number of edges in the bilayer networks, such that the third example in Fig. 4 would have the largest overlap value with itself.

Nevertheless, we could make use of the "aristocratic" approach to the backbone searching problem as described in [29] and only compare those configurations with each other which have large overlap values with each other. Here we would need to study whether all these configurations share a large backbone in the way that some subgraph of the bilayer network is always identical and could be used to perform a gradual reaction scheme for e.g. generating a desired macromolecule. Then one would simply fill all other droplets with e.g. water only, such that they would not disturb the overall reaction processes.

However, it could also turn out that such common properties are not commonly shared among the configurations but that some properties turn up with some larger probabilities, as found for dense packings of multidisperse systems of hard discs [30]. We will see what the outcome of our investigations will be.

6   Ultrametricity

Complex systems often exhibit the property of ultrametricity in configuration space or at least in a subspace of quasi-optimum configurations. As it was also found for a related hard disc packing problem [31], we expect this property also to turn up for this problem.

In order to understand the term ultrametricity, let us first refer to a standard metric. A standard metric like the Euclidean metric is defined as follows:

- All distances between pairs (i, j) of nodes are nonnegative, i.e., $d(i, j) \geq 0$ $\forall (i, j)$.
- The distance of a node to itself is $d(i, i) = 0$.
- In the case that a metric is symmetric, we have $d(i, j) = d(j, i)$ for all pairs of nodes.
- The most important property of a metric is the triangle inequality: $d(i, j) \leq d(i, k) + d(k, j)$ for all triples (i, j, k) of nodes, i.e., making a detour via a third node k can never be shorter than traveling directly from i to j.

For an ultrametric, the triangle inequality is replaced by the condition

$$d(i, j) \leq \max\{d(i, k), d(k, j)\} \tag{5}$$

for all triples of nodes (i, j, k).

If permuting i, j, k and applying the ultrametricity condition to all permutations, one finds that the nodes have to be placed on triangles which are either equilateral or at least isosceles with short base line, if the ultrametricity condition is fulfilled.

The question is now how to derive distances between bilayer networks. Here we can make use of the overlaps we defined earlier. By normalizing the overlap to

$$q_{\sigma\tau} = Q(\sigma, \tau)/Q_{\max}, \tag{6}$$

we can e.g. define a distance by

$$d(\sigma, \tau) = 1 - q_{\sigma\tau}, \tag{7}$$

i.e., the larger the overlap is, the smaller is the distance between the corresponding configurations. As demonstrated in [31], the existence of ultrametricity can then be proven by having a look at the joint probability distribution $p(d(\sigma, v), d(v, \tau))$ for various fixed values of $d(\sigma, \tau)$. If this joint probability distribution exhibits when plotted a significant peak along the diagonal and if this signal remains also if subtracting the product of the probabilities for the corresponding distance values, i.e., if considering $p(d(\sigma, v), d(v, \tau)) - p(d(\sigma, v)) \times p(d(v, \tau))$, then we find that the system really exhibits the property of ultrametricity.

If this property is fulfilled, then one knows that the resulting three-dimensional arrangements can be gathered in groups, which can be combined to supergroups, those again to hypergroups, and so on. By this successive gathering process, we basically generate an ultrametric tree, which is also well known from other fields of research, like the problem of constructing a phylogenetic tree. In order to construct such a tree, we can make use of the neighbor-joining method, which is a standard tool to reconstruct phylogenetic trees [32,33]. Another possibility would be to generate a clustered ordering of configurations [34].

7   Replica Symmetry Breaking

Ultrametricity in turn is related to iterated replica symmetry breaking (RSB). In order to derive RSB, let us start off with the replica symmetric assumption. As a simplification, it is assumed that the matrix $(q_{\sigma\tau})$ of normalized overlap values has the form

$$
(q_{\sigma\tau}) =
\begin{pmatrix}
1 & q & q & q & q & q & q & q & q & q & q & q \\
q & 1 & q & q & q & q & q & q & q & q & q & q \\
q & q & 1 & q & q & q & q & q & q & q & q & q \\
q & q & q & 1 & q & q & q & q & q & q & q & q \\
q & q & q & q & 1 & q & q & q & q & q & q & q \\
q & q & q & q & q & 1 & q & q & q & q & q & q \\
q & q & q & q & q & q & 1 & q & q & q & q & q \\
q & q & q & q & q & q & q & 1 & q & q & q & q \\
q & q & q & q & q & q & q & q & 1 & q & q & q \\
q & q & q & q & q & q & q & q & q & 1 & q & q \\
q & q & q & q & q & q & q & q & q & q & 1 & q \\
q & q & q & q & q & q & q & q & q & q & q & 1
\end{pmatrix},
\tag{8}
$$

i.e., all overlap values between different configurations are equal to one value q with $0 < q < 1$. This concept of replica symmetry (RS) is often used in theoretical physics as a first step to approximately calculate some properties [35].

In the next step, one allows for a first breaking of the replica-symmetry (RSB1), such that the overlap matrix e.g. looks like

$$
(q_{\sigma\tau}) =
\left(
\begin{array}{ccc|ccc|ccc|ccc}
1 & q_0 & q_0 & q_1 & q_1 & q_1 & q_1 & q_1 & q_1 & q_1 & q_1 & q_1 \\
q_0 & 1 & q_0 & q_1 & q_1 & q_1 & q_1 & q_1 & q_1 & q_1 & q_1 & q_1 \\
q_0 & q_0 & 1 & q_1 & q_1 & q_1 & q_1 & q_1 & q_1 & q_1 & q_1 & q_1 \\
\hline
q_1 & q_1 & q_1 & 1 & q_0 & q_0 & q_1 & q_1 & q_1 & q_1 & q_1 & q_1 \\
q_1 & q_1 & q_1 & q_0 & 1 & q_0 & q_1 & q_1 & q_1 & q_1 & q_1 & q_1 \\
q_1 & q_1 & q_1 & q_0 & q_0 & 1 & q_1 & q_1 & q_1 & q_1 & q_1 & q_1 \\
\hline
q_1 & q_1 & q_1 & q_1 & q_1 & q_1 & 1 & q_0 & q_0 & q_1 & q_1 & q_1 \\
q_1 & q_1 & q_1 & q_1 & q_1 & q_1 & q_0 & 1 & q_0 & q_1 & q_1 & q_1 \\
q_1 & q_1 & q_1 & q_1 & q_1 & q_1 & q_0 & q_0 & 1 & q_1 & q_1 & q_1 \\
\hline
q_1 & q_1 & q_1 & q_1 & q_1 & q_1 & q_1 & q_1 & q_1 & 1 & q_0 & q_0 \\
q_1 & q_1 & q_1 & q_1 & q_1 & q_1 & q_1 & q_1 & q_1 & q_0 & 1 & q_0 \\
q_1 & q_1 & q_1 & q_1 & q_1 & q_1 & q_1 & q_1 & q_1 & q_0 & q_0 & 1
\end{array}
\right),
\tag{9}
$$

with two overlap values $0 < q_1 < q_0 < 1$. Thus, we find block matrices with a larger overlap value along the diagonal of the matrix. In the second replica-symmetry breaking step (RSB2), the overlap matrix would e.g. look like

$$
(q_{\sigma\tau}) = \left(
\begin{array}{ccc|ccc||ccc|ccc}
1 & q_0 & q_0 & q_1 & q_1 & q_1 & q_2 & q_2 & q_2 & q_2 & q_2 & q_2 \\
q_0 & 1 & q_0 & q_1 & q_1 & q_1 & q_2 & q_2 & q_2 & q_2 & q_2 & q_2 \\
q_0 & q_0 & 1 & q_1 & q_1 & q_1 & q_2 & q_2 & q_2 & q_2 & q_2 & q_2 \\
\hline
q_1 & q_1 & q_1 & 1 & q_0 & q_0 & q_2 & q_2 & q_2 & q_2 & q_2 & q_2 \\
q_1 & q_1 & q_1 & q_0 & 1 & q_0 & q_2 & q_2 & q_2 & q_2 & q_2 & q_2 \\
q_1 & q_1 & q_1 & q_0 & q_0 & 1 & q_2 & q_2 & q_2 & q_2 & q_2 & q_2 \\
\hline\hline
q_2 & q_2 & q_2 & q_2 & q_2 & q_2 & 1 & q_0 & q_0 & q_1 & q_1 & q_1 \\
q_2 & q_2 & q_2 & q_2 & q_2 & q_2 & q_0 & 1 & q_0 & q_1 & q_1 & q_1 \\
q_2 & q_2 & q_2 & q_2 & q_2 & q_2 & q_0 & q_0 & 1 & q_1 & q_1 & q_1 \\
\hline
q_2 & q_2 & q_2 & q_2 & q_2 & q_2 & q_1 & q_1 & q_1 & 1 & q_0 & q_0 \\
q_2 & q_2 & q_2 & q_2 & q_2 & q_2 & q_1 & q_1 & q_1 & q_0 & 1 & q_0 \\
q_2 & q_2 & q_2 & q_2 & q_2 & q_2 & q_1 & q_1 & q_1 & q_0 & q_0 & 1 \\
\end{array}
\right), \tag{10}
$$

with three overlap values $0 < q_2 < q_1 < q_0 < 1$. Thus, the block matrices along the diagonal are enclosed in even larger block matrices. This approach can be iterated ad infinitum. Using this approach, Parisi was able to find the minimum energy for the Sherrington-Kirkpatrick spin glass model [36,37]. By using the clustering approach as described in [34], we could also try to find an ordering among the various different three-dimensional arrangements of droplets, such that an overlap matrix could look this way. Then we would know that we have groups of resulting bilayer configurations which have many properties in common or that we even have groups of configurations which allow for a small number of parallel evolving successive chemical reaction schemes.

8   Summary and Outlook

Within the European Horizon 2020 project *ACDC*, we aim at the development of a chemical compiler being able to govern biochemical reactions in a cheap and portable biochemical laboratory, intended to e.g. create specific macromolecules, like antibiotics. For this purpose, a microfluidic system is used in which droplets are generated which arrange themselves in a three-dimensional way and form bilayers with neighboring droplets. These bilayers form a network which can be used for some specific successive biochemical reaction scheme. Besides trying to understand and to simulate this spatial transition [4] in order to foretell the experimental outcome, we also have to investigate the final droplet arrangements and their corresponding bilayer networks, achieved in experiments and from computer simulations, in order to later on be able to design experiments in a way that some specific bilayer networks are created being able to produce the macromolecules desired.

In this paper, we have dwelt on thoughts about how to characterize differences between various three-dimensional agglomerations of droplets and their corresponding bilayer networks. If there are only small and few differences, searching for a network core might be a suitable strategy. Then all the configurations exhibit the same core network, which can then be used for the reaction scheme instead of the overall network. If this strategy is not successful, we can use the Searching for Backbones algorithm in order to detect parts which are common to all configurations and then determine whether it is possible to use the various ways in which these common parts are connected in the various bilayer networks in order to create the desired macromolecules. But if comparing all configurations in parallel, one might overlook that there might be groups consisting of configurations which are rather similar to each other within each group but exhibit larger differences to configurations in other groups. It might even be that not only various configurations can be merged to groups but that we can iterate this strategy, thus finding supergroups and hypergroups. For various complex problems, such successive mergings of configurations have already been detected by investigating ultrametric properties and replica symmetry breaking, which we also intend to use.

If we have been able to create configurations in Monte Carlo simulations similar to those found in experiments and even to foretell which configurations will be created if changing the experimental situation as described in [4], and if we have achieved this second part of our objective of understanding and predicting the outcome of an experimental setup, i.e., the various groups of three-dimensional arrangements of droplets generated, then we will be able to create a probabilistic chemical compiler in the final stage of this project. We aim at creating plans for e.g. a step-wise generation of some desired macromolecules, which are gradually constructed from smaller units, being contained in the various droplets, with the successive chemical reactions being enabled via the bilayers formed between neighboring droplets. Such a compiler has been exemplarily already developed for one specific molecule [2]. In this project, this compiler has to be generalized and also made probabilistic because of the variability in the rearrangement process which is to be expected.

Acknowledgment. JJS is deeply thankful to Uwe Krey, University of Regensburg, Germany, for teaching him the foundations of ultrametricity and replica symmetry breaking. Furthermore, JJS would like to kindly acknowledge fruitful discussions on this subject with Giorgio Parisi at the Sapienza University of Rome, Italy, with Scott Kirkpatrick at The Hebrew University of Jerusalem in Israel, with Ingo Morgenstern at the University of Regensburg, Germany, and with Peter van Dongen at the Johannes Gutenberg University of Mainz, Germany.

This work has been financially supported by the European Horizon 2020 project *ACDC – Artificial Cells with Distributed Cores to Decipher Protein Function* under project number 824060.

References

1. Li, J., Barrow, D.A.: A new droplet-forming fluidic junction for the generation of highly compartmentalised capsules. Lab Chip **17**, 2873–2881 (2017)

2. Weyland, M.S., et al.: The MATCHIT automaton: exploiting compartmentalization for the synthesis of branched polymers. Comput. Math. Methods Med. **2013**, 467428 (2013)
3. Flumini, D., Weyland, M.S., Schneider, J.J., Fellermann, H., Füchslin, R.M.: Steps towards programmable chemistries. Accepted for publication in the Wivace 2019 conference proceedings. In: XIV International Workshop on Artificial Life and Evolutionary Computation, Rende, Italy, 18–20 September 2019
4. Schneider, J.J., Weyland, M.S., Flumini, D., Matuttis, H.-G., Morgenstern, I., Füchslin, R.M.: Studying and simulating the three-dimensional arrangement of droplets. Accepted for publication in the Wivace 2019 conference proceedings. In: XIV International Workshop on Artificial Life and Evolutionary Computation, Rende, Italy, 18–20 September 2019
5. Morgan, A.J.L., et al.: Simple and versatile 3D printed microfluidics using fused filament fabrication. PLoS ONE **11**(4), e0152023 (2016)
6. Eggers, J.: Nonlinear dynamics and breakup of free-surface flows. Rev. Mod. Phys. **69**, 865 (1997)
7. Eggers, J., Villermaux, E.: Physics of liquid jets. Rep. Progress Phys. **71**, 036601 (2008)
8. Link, D.R., Anna, S.L., Weitz, D.A., Stone, H.A.: Geometrically mediated breakup of drops in microfluidic devices. Phys. Rev. Lett. **92**, 054503 (2004)
9. Garstecki, P., Stone, H.A., Whitesides, G.M.: Mechanism for flow-rate controlled breakup in confined geometries: a route to monodisperse emulsions. Phys. Rev. Lett. **94**, 164501 (2005)
10. Garstecki, P., Fuerstman, M.J., Stone, H.A., Whitesides, G.M.: Formation of droplets and bubbles in a microfluidic T-junction - scaling and mechanism of breakup. Lab Chip **6**, 437–446 (2006)
11. Guillot, P., Colin, A., Ajdari, A.: Stability of a jet in confined pressure-driven biphasic flows at low Reynolds number in various geometries. Phys. Rev. E **78**, 016307 (2008)
12. Au, A.K., Huynh, W., Horowitz, L.F., Folch, A.: 3D-printed microfluidics. Angew. Chem. Int. Ed. **55**, 3862–3881 (2016)
13. Takenaga, S., et al.: Fabrication of biocompatible lab-on-chip devices for biomedical applications by means of a 3D-printing process. Phys. Status Solidi **212**, 1347–1352 (2015)
14. Macdonald, N.P., Cabot, J.M., Smejkal, P., Guit, R.M., Paull, B., Breadmore, M.C.: Comparing microfluidic performance of three-dimensional (3D) printing platforms. Anal. Chem. **89**, 3858–3866 (2017)
15. Lee, K.G., et al.: 3D printed modules for integrated microfluidic devices. RSC Adv. **4**, 32876–32880 (2014)
16. Yazdi, A.A., Popma, A., Wong, W., Nguyen, T., Pan, Y., Xu, J.: 3D Printing: an emerging tool for novel microfluidics and lab-on-a-chip applications. Microfluid. Nanofluid **20**, 1–18 (2016)
17. Chen, C., Mehl, B.T., Munshi, A.S., Townsend, A.D., Spence, D.M., Martin, R.S.: 3D-printed microfluidic devices: fabrication, advantages and limitations - a mini review. Anal. Methods **8**, 6005–6012 (2016)
18. He, Y., Wu, Y., Fu, J., Gao, Q., Qiu, J.: Developments of 3D printing microfluidics and applications in chemistry and biology: a review. Electroanalysis **28**, 1–22 (2016)
19. Tasoglu, S., Folch, A. (eds.): 3D Printed Microfluidic Devices. MDPI, Basel (2018)
20. Li, J.: Private communication (2019)

21. Babai, L.: Graph Isomorphism in Quasipolynomial Time (2016). https://arxiv.org/pdf/1512.03547.pdf
22. Jensen, T.R., Toft, B.: Graph Coloring Problems. Wiley, New York (1995)
23. Ringel, G., Youngs, J.W.T.: Solution of the Heawood map-coloring problem. Proc. Nat. Acad. Sci. **60**, 438–445 (1968)
24. Carmi, S., Havlin, S., Kirkpatrick, S., Shavitt, Y., Shir, E.: A model of Internet topology using k-shell decomposition. PNAS **104**, 11150–11154 (2007)
25. Schneider, J., Froschhammer, C., Morgenstern, I., Husslein, T., Singer, J.M.: Searching for backbones - an efficient parallel algorithm for the traveling salesman problem. Comp. Phys. Comm. **96**, 173–188 (1996)
26. Schneider, J.: Searching for backbones - a high-performance parallel algorithm for solving combinatorial optimization problems. Electron. Notes Future Generation Comput. Syst. **1**, 121–131 (2001)
27. Schneider, J.: Searching for Backbones - a high-performance parallel algorithm for solving combinatorial optimization problems. Future Generation Comput. Syst. **19**, 121–131 (2003)
28. Schneider, J.J.: Searching for Backbones - an efficient parallel algorithm for finding groundstates in spin glass models. In: Tokuyama, M., Oppenheim, I. (eds.) 3rd International Symposium on Slow Dynamics in Complex Systems, Sendai, Japan. AIP Conference Proceedings, vol. 708, pp. 426–429 (2004)
29. Schneider, J.J., Kirkpatrick, S.: Stochastic Optimization. Springer, Heidelberg, New York (2006). https://doi.org/10.1007/978-3-540-34560-2
30. Müller, A., Schneider, J.J., Schömer, E.: Packing a multidisperse system of hard disks in a circular environment. Phys. Rev. E **79**, 021102 (2009)
31. Schneider, J.J., Müller, A., Schömer, E.: Ultrametricity property of energy landscapes of multidisperse packing problems. Phys. Rev. E **79**, 031122 (2009)
32. Saitou, N., Nei, M.: The neighbor-joining method: a new method for reconstructing phylogenetic trees. Mol. Biol. Evol. **4**, 406–425 (1987)
33. Studier, J.A., Keppler, K.J.: A note on the neighbor-joining algorithm of Saitou and Nei. Mol. Biol. Evol. **5**, 729–731 (1988)
34. Schneider, J.J., Bukur, T., Krause, A.: Traveling salesman problem with clustering. J. Stat. Phys. **141**, 767–784 (2010)
35. Sherrington, D., Kirkpatrick, S.: Solvable model of a spin glass. Phys. Rev. Lett. **35**, 1792–1796 (1975)
36. Parisi, G.: Infinite number of order parameters for spin-glasses. Phys. Rev. Lett. **43**, 1754–1756 (1979)
37. Parisi, G.: A sequence of approximate solutions to the S-K model for spin glasses. J. Phys. A **13**, L-115 (1980)

Correction to: Artificial Life and Evolutionary Computation

Franco Cicirelli, Antonio Guerrieri, Clara Pizzuti,
Annalisa Socievole, Giandomenico Spezzano,
and Andrea Vinci

Correction to:
F. Cicirelli et al. (Eds.): *Artificial Life and Evolutionary*
Computation, **CCIS 1200,**
https://doi.org/10.1007/978-3-030-45016-8

The following chapters were originally published electronically on the publisher's internet portal without open access:

"Towards Programmable Chemistries", written by Dandolo Flumini, Mathias S. Weyland, Johannes J. Schneider, Harold Fellermann, Rudolf M. Füchslin;

"Studying and Simulating the Three-Dimensional Arrangement of Droplets", written by Johannes Joscf Schneider, Mathias Sebastian Weyland, Dandolo Flumini, Hans-Georg Matuttis, Ingo Morgenstern, Rudolf Marcel Füchslin;

"Investigating Three-Dimensional Arrangements of Droplets", written by Johannes Josef Schneider, Mathias Sebastian Weyland, Dandolo Flumini,Rudolf Marcel Füchslin.

With the authors' decision to opt for Open Choice the copyright of the chapters changed on 19 September 2023 to © Authors, 2023 and the chapters are forthwith distributed under a Creative Commons Attribution.

The updated original version of these chapters can be found at
https://doi.org/10.1007/978-3-030-45016-8_15
https://doi.org/10.1007/978-3-030-45016-8_16
https://doi.org/10.1007/978-3-030-45016-8_17

Funded by: the European Union's Horizon 2020 program "Artificial Cells with Distributed Cores to Decipher Protein Function" (ACDC), Grant Number: 824060.

Author Index

Printed in the United States
by Baker & Taylor Publisher Services